E. Andrew Boyd presents us with a wonderful ride through some of the great riddles of the ages. He does so to elaborate his "Beyond Comprehension" conjecture—that realities exist beyond our hope of ever comprehending them. That conjecture is at once obvious, while it screams out for our denial. And the very fact that it cannot be proved either way only adds to its validity. All the while we have his examples—Oh the fine mind-teasing examples!

Dr. John Lienhard
National Academy of Engineering
Creator, The Engines of Our Ingenuity
M.D. Anderson Professor of Technology and Culture, Emeritus
University of Houston

When I was young, I was drawn to mathematics by concepts that were beyond ordinary comprehension. Boyd's book explains limits to human comprehension in a way that is accessible to a broad audience. In the spirit of *Freakonomics,* and books by Malcolm Gladwell, Bill Bryson, and Neil deGrasse Tyson, Boyd inspires a sense of wonder about the outer limits of comprehension in mathematics and physics.

Dr. L. Ridgway Scott
Louis Block Professor of Mathematics and Computer Science
University of Chicago

Beyond Comprehension

A Scientific Look at the
Challenge of Knowing Everything

E. Andrew Boyd

To Rick —
Happy Reading!
— Andy Boyd

HAMILTON-HAVERBROOK
Worldwide

Beyond Comprehension

A Scientific Look at the Challenge of Knowing Everything

For information regarding permission to reproduce selections from this book, contact
hamilton-haverbrook@gmail.com

First edition

1. Science—Philosophy 2. Mathematics—Philosophy—Foundations 3. Paradox 4. Infinity
5. Absurdity 6. Consciousness 7. Enigma

ISBN 978-0-9992087-0-0 pbk.

HAMILTON-HAVERBROOK
Worldwide

for Alex

CONTENTS

Acknowledgments

If memory would allow I'd like to thank all of the people who've contributed to this book, which is to say anyone I've ever had a mathematical, scientific or philosophical discussion with, not to mention those individuals I've only had the chance to encounter through their writings. As my memory isn't up to the task, I'll have to stick a little closer to home.

David Wood was a terrific help in reviewing materials, assisting with some filming related to the book, and through regular discussions. The effervescent mind of Paul Yost was always a source of inspiration and good humor. Copy editor Erica Smith did a terrific job.

Most of all I'd like to thank my wife, Sarah, who's put up with what I'm sure were seemingly endless scientific and philosophical discussions. As she's a professor of French history I'm aware there are topics of more interest to her, but she good-heartedly lent an ear, offered ideas, and productively challenged my thinking—as she has for so many things, for so many years.

Preface

A teacher stands before a classroom of students who are busily taking notes.

"And in my view, Jefferson's defense of these basic rights lacked conviction. Any discussion?"

Silence. The teacher continues.

"Let me just add that personally I believe the Bill of Rights to be a silly, inconsequential recapitulation of truths already found in the Constitution. Any comment?"

More silence.

"No, *scratch* that! The Constitution *itself* should never have been ratified! It's a dangerous document! All power should rest with the *executive!* What do you think of *that!*"

Even more silence.

"JEFFERSON WAS THE *ANTICHRIST!* DEMOCRACY IS FASCISM! BLACK IS WHITE! NIGHT IS DAY!"

"Boy, this course is getting interesting," says one student to another.

"You said it," comes the response. "I didn't know half this stuff."

This *Doonesbury* comic strip first appeared in 1985 and spoke to sixties-educated teachers who were confronting a new generation of students; a generation that seemed more interested in grades than in changing the world. How could students be so blind and complacent that equating black and white failed to evoke incredulity?

As humankind charges headlong into the future, I sometimes feel as though we, too, are spending so much time taking notes that we're failing to appreciate the world we're confronted with. Throughout most of human history the universe was filled with mystery. Where did the sun go at night? Why did humankind hold a special place in the universe? With advances in physics, biology and the other sciences, the world's certainly become a much less mysterious place. Yet, as we probe ever more deeply, we find we're still facing puzzles that defy explanation. We laugh, make note of them, and file them away in the recesses of our brains. But we shouldn't. We should marvel at them. In *Beyond Comprehension* I dust off some of life's special puzzles in an effort to evoke a combination of incredulity and astonishment.

None of the topics we'll look at are altogether new. In fact, many have entire books written about them, and I encourage readers who encounter something new and exciting to follow up with one of the references I've provided. My intent isn't to delve too deeply into any one topic but to share the wonder found in many.

To achieve this sense of wonder requires help. All of the topics have been chosen for their enigmatic nature. But while some tend to jump off the page, others require reflection. In some cases the biggest challenge may be familiarity. For example, I've observed that many people have trouble getting excited about gravity. It's such a part of our daily lives that it takes effort to break free of our mind-set and appreciate how remarkable gravity really is and why Newton's theory of gravity was once viewed with great skepticism.

Another challenge is the natural human desire to explain things. Explanations are important in that they help as we try to make sense of something new that we encounter. But explanations aren't a requirement, especially when they're not forthcoming or miss the fundamental point. It's okay to simply stop and smell the roses.

A few housekeeping items. All chapters stand independently except for "Reflections," which is why I've included all notes, references, and appendices with each chapter rather than at the end of the book.

Also, I've done my best to use as little math as possible, but I couldn't bring myself to shy away from it altogether. For those who are interested, the mathematical details provide another level at which to marvel over the perplexities of our world. If in the course of reading you encounter something technical that doesn't interest you, skip it. Doing so shouldn't hamper your progress. For the record, I've used nothing beyond high school math in the pages that follow, and most of the time nothing beyond elementary school math. Reading the first and last chapters is also an option, returning to the intermediate chapters when time and interest allow.

I wish you a happily befuddling journey on your road to discovery.

E. Andrew Boyd

It is of great use to the sailor to know the length of his line,
though he cannot with it fathom all the depths of the ocean.

– John Locke

CHAPTER ONE

No Mystery:
The Limits of Human Comprehension

Dogs are remarkable creatures. They display emotion when playing or cuddling with their owners. They form complex relationships with people and other animals. And they're keenly intelligent, capable of learning hundreds of words, from "sit" to "fetch." But there's something pretty basic dogs don't grasp.

The number ten.

And they're not alone. Ten is a foreign concept to all animals other than humans. Research has shown that many animals are able to discern when one set of things is larger than another, and some animals demonstrate a distinct knowledge of the numbers one, two, and three. But go much beyond that and the animal mind withers to the task. It's not that the dog mind doesn't understand the concept of ten; that it somehow recognizes the digits on two human hands as "something" but doesn't know what to do with that "something." The specific concept simply doesn't materialize. The dog brain isn't wired to wrap itself around "ten." Dogs and other animals don't fail to understand ten. Ten is *beyond their comprehension.*

1

Yet for humans the number ten is so elementary we take it for granted. Ten comes after nine and before eleven. Two fives make ten. Ten plus two equals twelve. All of these simple observations—much less the edifice of mathematics that underlies so much of our modern technology—are completely beyond the mental capacity of animals.

Mathematics is just one realm where the human mind comprehends what the animal mind cannot. You and I might agree to meet downtown at a coffee shop next Tuesday at 7:00 a.m., and barring forgetfulness or an emergency, we'll drive our cars and get together at the appointed time. The animal mind would be confounded on many fronts, from understanding the rules of the road to the task of establishing a specific meeting time. Again, the issue is not a lack of understanding but a lack of comprehension. It seems some animals have a vague sense of time in the same way they have a vague sense of bigger and smaller numbers, but even that's open to debate. Many researchers believe animals truly "live in the moment"—that they have no sense of past or future.

Yet time comes so naturally to us that we rarely give it a second thought. And we certainly don't live our lives in the moment. We remember lessons from the past and plan for the future—often causing ourselves unnecessary stress in the process.

The point is not to dwell on the cognitive prowess of humans, though the human mind, stemming from the human brain, is extraordinary when compared to that of other animals. The point is simply to provide background for the following widely accepted observation:

Observation: *There are real things in our world that dogs and other animals can't comprehend.*

It's worth emphasizing two very important points. First, it can't be overstated that *can't comprehend* is different from a mere lack of understanding. I may not understand how an automobile engine works, but if I took enough time, I could learn. And even without knowing the details, I know that when gasoline burns, it releases energy, which the engine harnesses in a way that makes a car move. I can comprehend what an engine is even if I don't fully understand it. To say that animals can't comprehend numbers means something much stronger. It means they don't even know what they're missing.

The second point of emphasis is the use of the term *real things*. We're not talking about something fanciful. The two examples just mentioned are the existence of numbers and the passage of time. Even if all human life were extinguished, the number ten would still be ten and time would keep marching forward. This is the sense in which we use the word "real."

Our observation sets the stage for the following conjecture:

Beyond Comprehension (BC) Conjecture: *There are real things in our world that humans can't comprehend.*

The foundation for this conjecture is quite simple: humans are animals, too. Why should we, as a species, expect that we're wired to comprehend everything? No other animals can. Has evolution singled us out, taking us to a pinnacle from which we're able to comprehend all that is? If anything, evolution suggests just the opposite. Every animal we look at, at least for the time being, has reached a cognitive plateau. We know they haven't reached a pinnacle since we can look at our own cognitive abilities and see that other animals still have a long way to go. Extrapolating, it seems far, far more likely that we, too, have reached a plateau and not a pinnacle.

Fully absorbing the conjecture takes some reflection. Our perspective on the world is through that which we can comprehend. How can there be real things we can't comprehend? Or even if there are, why should we care? After all, they're beyond our comprehension, which means we'll never be able to wrap our heads around them in any meaningful sense. Whether they're real or not, since we can't comprehend them, they're irrelevant—at least from the perspective of rational, scientifically minded humans.

Setting aside the question of why we should care for the final chapter of this book, there may be a way to infer the existence of real things we can't comprehend. But how do we go about this? To begin our journey, we acknowledge that while we can't comprehend what we can't comprehend, we may still be able to see shadows of the incomprehensible. Such shadows might present themselves as paradoxes—inexplicable realities. Or we may find ourselves staring at something so befuddling that our only recourse is to succumb to its incomprehensibility. This, of course, doesn't constitute a proof of our conjecture. But coupled with our observation about the limitations of dogs and other animals, it provides evidence. And the examples we'll draw upon from mathematics, physics, philosophy, neuroscience, and computer science provide a glimpse into what a wonderfully surprising world we live in. But before we set off in search of things that can't be, let's dwell for a moment on human comprehension.

*****

Adam is special. He doesn't drink, smoke, or do drugs out of a personal conviction they're bad. He's diligent with laundry and happy to help out when something needs doing around the house. He's strong, healthy, and stays in good shape by watching his weight and diet. Adam faithfully attends church on Sundays and remembers his deceased relatives with great affection—especially a grandfather he was especially close to. In many ways, he's a parent's dream.

But Adam is special in other ways, too, as people discover when they spend a little time with him. A recent conversation with Adam went as follows.

> **Adam:** People wear towels on their heads when it's hot outside.
>
> **Family member:** Yes, some people dip them in water to keep cool. It also helps keep the sun off their necks and faces.

Adam: They don't wear hats, they wear towels.

Family member: Sometimes they wear hats, too.

Adam: Most wear towels.

Family member: Okay. How's your day going?

Adam: I wear a towel.

Family member: I've noticed you doing that lately. Work go okay?

Adam: Other people wear towels when they work in the sun.

Family member: Yes. Did you do your exercise this morning?

Adam: The towel helps me keep cool.

Adam has autism. People who chance to meet him often feel sad about his limited mental capacity. But those who've spent any length of time with him look at it from a different perspective. It's not that he can't think straight. He just thinks differently.

Differently comes in many forms. Adam's conversational abilities and his understanding of the nuances of language are deficient by most measures. When Adam was young he had great difficulty with personal pronouns—"I" and "you" in particular. The reason, which quickly became clear, was that they change with who's using them. When I say "I," I'm referring to myself. When you say "I," you're referring to yourself. "I" changes depending on who utters it. That's not true of a name. I say Adam, you say Adam, we all say Adam. Adam made mistakes for years before finally mastering personal pronouns.

He also assumes others can make connections that they don't, or at least that aren't obvious. As a child of four, Adam proclaimed to his parents, "I hate the babies." He'd always shown affection for other children, large and small, and his parents were at a loss as to why Adam had concluded that he hated the babies. Adam continued to express his disenchantment in the days that followed. Finally, out of the blue, he added, "Babies poop." His parents immediately made the connection.

Adam had been going to the community pool daily, and whenever waste from a diaper was found floating in the pool, it was closed for super chlorinating, thus spoiling Adam's fun. Adam didn't hate babies. He hated the fact that babies caused the pool to close.

It was the beginning of a lifetime in which Adam presumed other minds were making connections they didn't. A particularly disturbing incident occurred when Adam was in his mid-twenties. "Police keep us safe," he said, and he continued on variants of the theme for days before changing to "police arrest burglars" and "bad people" and finally to "police use handcuffs" and "put burglars in their cars." At this point, his parents questioned him more directly and the full story came out. While looking for a burglar, police ran across Adam. Given his cryptic responses and lack of physical anomalies, they searched him, hand-cuffed him, and put him in the back of their patrol car. More details weren't forthcoming, except for the fact that the story ended with the police dropping him off at his home and explaining that they were look-ing for a burglar (presumably to keep Adam and his neighbors safe). From Adam's first mention of police to the unraveling of the story took a full week.

Adam clearly lacks cognitive skills in areas that most people take for granted. But in other areas, Adam demonstrates uncommon abilities. He remembers the days and dates of obscure events, such as when an aunt received a traffic ticket years in the past. When events have been checked, his memories have proven accurate.

Perhaps most striking was a feat of mathematical gymnastics Adam performed at age 14, when he informed his father they were "reversed ages." Adam's father was 41 at the time and surmised his son was refer-ring to the digits being reversed—one-four, four-one. Adam went on to point out that they would be reversed ages again when he was 25 and the father 52, and then again when he was 36 and the father 63. It was Adam's way of saying reversed digits repeated every 11 years. That's not especially surprising since adding 11 to any number is the same as adding one to each digit no matter what the order; 14 plus 11 equals 25, and 41 plus 11 equals 52. More interestingly, Adam went on to point out other relatives who would share a reversed age with him and when, and he pointed out that others never would.

All these facts were easily verified by his family, but they raised a question. Is there a simple mathematical property that makes some people have reversed ages while others don't? The answer wasn't obvious at first glance. So after a few days of thought they asked Adam: what makes people have reversed ages? His reply: "multiples of 9." When asked again, he didn't elaborate but simply reiterated what he'd said before: "multiples of 9." Interestingly, neither 14 nor 41 is a multiple of 9, but their difference, 27, is. It takes a little effort to prove, but it's exactly the condition that leads to reversed ages. Adam and his father periodically share reversed ages with each other because their age difference is a multiple of 9. Anyone whose age difference with Adam isn't a multiple of 9 will never share reversed ages with him.

It doesn't require advanced training in neurobiology to realize Adam comprehends the world dissimilarly to most people. His brain is wired differently, giving rise to its own strengths and weaknesses—its own set of cognitive limitations.

Adam's condition causes his cognitive limitations to stand out. But are "normal" people really all that normal? Apart from minor differences, do our brains have the same basic limitations?

The question came alive for me when I was a junior in college. A math major, I earned money tutoring students who were having trouble with math. Most of the students I worked with were pursuing studies in the social sciences, where some familiarity with statistics was a prerequisite for obtaining a degree. One student in particular was having an extremely difficult time. She had no idea how to use the formulas she'd been given or even which formulas should be applied. She was a conscientious student who wanted to learn, and together we revisited the concepts behind the formulas. We worked diligently for weeks, but she continually failed to demonstrate even a basic understanding of the underlying abstractions. Could her mind simply have been unable to grasp them?

My sentiment was in clear contradiction to what I'd learned from my probability professor earlier in the year. "Anyone," he proclaimed, "can learn math. You just have to take the time and make the effort." To those for whom math comes easily, that's certainly true. In fact, for the mathematically skilled, it's difficult to understand why so many people seem to have perpetual difficulty with math. The abstractions are clear and precise—far more so than in any other discipline. They fit together like elegantly carved pieces in a jigsaw puzzle. Math isn't hard, it's a source of pleasure. But are the many underachievers simply failing to work hard enough, or are their minds simply incapable of comprehending higher math? Perhaps their brains are wired in such a way that no matter how much time and effort they put in, they'll never learn math beyond a certain level.

Which leads us to ask: how do we learn in the first place? In particular, do we learn everything from experience, or are there principles independent of experience that shape how we learn? Is the mind/brain a blank slate, a tabula rasa, or are we born with capabilities that help us acquire knowledge? Is the *empiricist* perspective that all knowledge comes from the senses correct, or is the *rationalist* position that we're born with the capacity to reason without relying on experience correct?

The question dates from as far back as ancient Greece and Plato. Plato's philosophy was deeply rationalist, proposing an ethereal world of ideal forms providing the standard against which objects in the material world were measured. Given the many different shapes and uses of tables, how do we recognize that the square wooden object in our kitchen where we eat and the small glass-and-metal object in the living room supporting a lamp are both tables? Plato answered this question by proposing that they're both shadows of an ideal table and that our minds have innate access to this ideal form. We recognize all tables as imperfect instances of the quintessential table that resides in an ethereal realm.

The rationalist/empiricist debate came to the fore, however, with the writings of René Descartes. A profound rationalist, Descartes pondered how it is that we comprehend wax. Imagine, he said, a piece of wax taken from a honeycomb. It's solid, it smells of honey and flowers, and its color, size, and shape are perfectly easy to ascertain. He then imagined setting the wax in front of the fire, causing it to melt. Everything changes, Descartes pointed out. The taste and smell vanish; the color, shape, and temperature change; the wax turns from solid to liquid. All the aspects by which we experience the wax through our senses change. Yet we still recognize that it's wax. Descartes concludes that our senses alone are insufficient to intelligibly comprehend the world around us. There must be something more—some inherent understanding of the world that doesn't depend on the senses. But what? Just as Plato had posited the existence of a world of ideal forms, Descartes and his rationalist contemporaries drew justification from a realm beyond the physical world. In Descartes's time, God stood forth as caretaker of that realm.

Such forays were increasingly met with rancor in the early seventeenth century, a period during which the Enlightenment was gaining momentum. Newton had shown how simple mathematical laws describe the working of the universe. This in turn led to an overall sense that the world was more machine than mystery. The human body and, by extension, the human mind weren't to be excluded from this new framework. It, too, was mechanical. And along with the mind-as-machine worldview came a new appreciation of the potential limits of human comprehension.

Empiricist John Locke challenged the views of Descartes and other rationalists on multiple fronts, believing that all knowledge is the result of "EXPERIENCE," where EXPERIENCE is a combination of "SENSATION" and "REFLECTION" (the capitalization is Locke's own). SENSATION consists of what our brains receive through our senses: taste, touch, smell, sight, and sound. REFLECTION is the operation of the human mind when it converts these sensations to ideas as diverse as "fire is hot" or "love brings happiness." An empiricist might explain our ability to discern what's a table and what isn't in the following way. We experience tables as we grow. Our parents tell us to take a seat at the dinner table or to straighten the coffee table. We then use reflection to abstract the various similarities (and differences from things that aren't tables) to arrive at an idea of what it means to be a table.[1]

Empiricism is typically thought of for its emphasis on the senses. But, more fundamentally, it was a challenge to the rationalist insistence on a mysterious spiritual domain. In empiricism we find an emphasis not only on sensory input, but on the fact that reflection is a brain process and therefore subject to the limitations of what the physical brain is capable of. Though Locke's primary purpose in his work *An Essay Concerning Human Understanding* was to focus on the senses in presenting his empiricist thesis, he was careful to explain that an important part of doing so was to understand the limits of human thought.

> For I thought that the first step towards satisfying several inquiries the mind of man was very apt to run into, was, to take a survey of our own understandings, examine our own powers, and see to what things they were adapted. Till that was done I suspected we began at the wrong end, and in vain sought for satisfaction in a quiet and sure possession of truths that most concerned us, whilst we let loose our thoughts into the vast ocean of Being; as if all that boundless extent were the natural and undoubted possession of our understandings, wherein there was nothing exempt from its decisions, or that escaped its comprehension.[2]

---

[1] Locke's position becomes especially clear when we consider the problem of describing a cat to a visitor from another planet. We might tell our visitor a cat is furry, with four legs, a tail, and whiskers. A dog then wanders by, and our visitor confidently, and incorrectly, proclaims, "cat." If, on the other hand, we'd shown the visitor a dozen cats and said, "cat," and a dozen dogs and said, "dog," he probably wouldn't have made the same mistake.

[2] John Locke, *An Essay Concerning Human Understanding, Volume 1, Books 1 and 2* (Project Gutenberg, 1690), www.gutenberg.org/ebooks/10615, Introduction, Section 7, *Occasion of this Essay*.

Locke's empiricism recognized that cognition was shaped not only by sensory experiences but by the workings of the mind/brain as interpreter of those experiences. A dog receives very similar sensory input to that of a human. It hears cars going by on the street, it smells the pot roast cooking on the stove, and every time its owner reaches out with a comforting hand it sees the familiar five digits we easily recognize as "five." But the dog's brain simply isn't structured to comprehend those five digits in the same way as a human's. Locke wanted to understand what the human mind was capable of before he sought to untangle life's great mysteries; he wanted to understand the length of the sailor's line before seeking to fathom the depths of the ocean.

While Descartes and Locke were fundamental in framing the rationalist/empiricist debate, they were by no means the only individuals involved in the discussion. Luminaries such as Baruch Spinoza, Gottfried Wilhelm Leibniz, George Berkeley, and David Hume—just to name a few—all contributed to the polemic. Immanuel Kant's *Critique of Pure Reason*, which offers an impressively detailed synthesis of the two positions, stands as one of the single greatest works in the history of philosophy. Today, most philosophers espouse some form of empiricism, in large part because empiricism aligns more easily with modern science than rationalism—at least as rationalism was conceived in Descartes's time.

While empiricism places emphasis on the senses and experience, the limitations of the mind/brain stemming from the body's physical nature were recognized at least as far back as the time of Locke. Only over time, however, has the study of mind/body as machine really become a central theme in scientific research.

Among those who contributed to the shift was the twentieth century philosopher Noam Chomsky. Chomsky studied language—in particular, how we learn to speak. Observing how quickly humans learn the relatively complex task of intricate verbal communication (something no other animals come close to) and recognizing the structural similarities of the many languages we speak, Chomsky was a proponent of *linguistic nativism*. As the name suggests, linguistic nativism advocates the position that humans are born with native abilities to acquire language. And not just any language. The similarities among languages are so great that they point to a prewired, biological basis that's relatively consistent among the many varied peoples of the world. No blank slate for Chomsky. Children learning a particular language are filling in the details, but the structure that allows the language to be learned is already imprinted in their brains.

Chomsky's position doesn't posit the existence of anything mysterious, as was the case with the early rationalists. Quite the contrary. He argues that the way we learn language is captured in the way our brains work. Among humans we may have difficulty communicating with people in other languages, but we're at least capable of learning those languages. Other brains, perhaps the brains of aliens that evolved on another planet, might have a very different structure that would lead to wholly different languages—languages so different it would be impossible for us to communicate with one another.

Chomsky argues that languages beyond our comprehension may well exist, and in doing so, he goes much further than many scientists and philosophers. Most agree the mind/brain is physical and is subject to physical limitations, but few step up to tackle the question of exactly what those limitations are. Chomsky's research has been widely recognized since, by focusing on language, he was able to put together cogent arguments defending his position. However, stronger claims that a particular question can't be answered simply because it's beyond our comprehension are typically, and appropriately, met with far more skepticism.

In recent years philosopher Colin McGinn has argued that many of the persistent questions in philosophy are beyond our comprehension and that just as we're prewired to understand language, we're most certainly not prewired to answer a host of questions about consciousness and self. McGinn refers to his position as *transcendental naturalism*, conveying the idea that certain philosophical questions transcend the basic, physical ability of our brains to intelligibly deal with them. It's also known by the name *cognitive closure*, and dismissively as the *new mysterianism*. As McGinn points out, the idea that the human mind/brain has limitations due to its physical nature isn't in itself controversial. The question is where we draw the line. And to many, McGinn has drawn the line where it doesn't belong, placing time-honored questions out of bounds. How can we simply write off the learned efforts of the great minds who've pondered consciousness and self, mind and brain? Or suggest we aren't ultimately capable of understanding some of the world's most captivating questions? Questions like how consciousness arises? Or, for that matter, how can we tell researchers in a mammoth industry—not only philosophy departments but also medical schools—that they're chasing answers to the unanswerable?

Philosopher Thomas Nagel has come under less criticism for his views on cognitive limitations of the human brain than McGinn, in large part because these limitations aren't at the core of Nagel's investigations. His interests are evident in his widely read paper, *What Is It Like to Be a Bat?*, in which he argues that the conscious experiences of different animals—humans and bats in particular—are different. His primary goal in the paper is to argue against reductionism—efforts to understand consciousness by deconstructing the physical workings of the brain. But along the way, he touches upon the limitations of the human mind/brain, stating that his philosophy "implies a belief in the existence of facts beyond the reach of human concept."

> Certainly it is possible for a human being to believe that there are facts which humans never *will* possess...But one might also believe that there are facts which *could* not ever be represented or comprehended by human beings, even if the species lasted

forever—simply because our structure does not permit us to operate with concepts of the requisite type.[3]

Like Nagel, philosopher John Searle doesn't raise eyebrows by pointing to specific lines of inquiry that are beyond comprehension. But he does argue that the failure to recognize our cognitive limitations has led people to wrong conclusions. In making his argument, Searle's position on the capabilities of the mind/brain becomes clear. And while most scientists and philosophers have never articulated their own positions on the limits of the mind/brain—or perhaps even contemplated them—most would probably agree with Searle.

> We should never forget who we are; and for such as us, it is a mistake to assume that everything that exists is comprehensible to our brains. Of course, methodologically we have to act as if we could understand everything, because there is no way of knowing what we can't: to know the limits of knowledge, we would have to know both sides of the limit. So potential omniscience is acceptable as a heuristic device, but it would be self-deception to suppose it a fact.[4]

Searle elevates the notion that there are real things in our world that humans can't comprehend from conjecture to statement of self-evident fact.

From autism to math, from Locke to Searle. It seems we may not be as omniscient as our natural human inclinations lead us to believe. But our goal in the following pages isn't to examine what these limitations might be. It's much more humble: to observe, and to confront situations that reason tells us just can't be. To puzzle over, chuckle about, and ultimately succumb to the fact that something, somehow, just isn't right. We are decidedly not, however, interested in mysteries.

---

[3]Thomas Nagel, "What Is It Like to Be a Bat?" in *Mortal Questions* (Cambridge: Cambridge University Press, 1979; first published 1974), 171.
[4]John Searle, *The Rediscovery of the Mind* (Cambridge, MA: MIT Press, 1992), 24.

*****

Comedian George Carlin was born an Irish Catholic, though he was quick to point out that by the time he'd grown up, he was an American. He attended his parish school in New York, a school that was traditional in its teaching of church doctrine but progressive in its educational methods: uniforms weren't required, no grades were given, and classes weren't segregated by sex. Unlike the dictatorial, ruler-toting environment Catholic schools were known for, the school created an environment of freedom. So much freedom, according to Carlin, that by eighth grade many of the students had lost the faith. Students were taught to raise questions—questions that sometimes evoked a response if not an answer: "It's a mystery."

Mystery is a loaded word. It can be used in many subtly different ways, but when it comes to expressing our understanding of something or lack thereof, it carries with it a sense of throwing in the towel. Take the statement "life's a mystery." The intent is to convey a sense that life is more than an organized collection of molecules capable of reproduction. Given our personal experience with our own lives, it's a justifiable belief. But the use of the word "mystery" evokes a sense that there's nothing more that can be said. We can seek to "experience" or "enter into" the mystery, but we can't hope to come to terms with it in any rational sense. Consider the following short-lived discussion between a scientist and a Christian theologian.

> **Scientist:** How do you explain the doctrine of the Trinity, that God, Jesus, and the Holy Spirit are both three distinct beings and one being at the same time?
> **Theologian:** It's a mystery.

In claiming the doctrine to be a mystery, the theologian has brought rational discussion to a halt.

By contrast, scientists avoid the word mystery because it implies something that can't be reckoned with through reason. The wave/particle duality of matter—where subatomic particles sometimes act like waves and at other times like particles—is as bewildering to physicists as the Trinity is to theologians. It's also perplexing, astounding, baffling, puzzling, head-scratching, confounding, and astonishing. But it's not a mystery, because the word mystery carries unwanted baggage: an implied "that's all, folks."

Mystery isn't a bad thing. Rocking a newborn. Watching the stars rise against a backdrop of remote mountains. Unexpected euphoria during a quiet moment. Each inspires a sense of mystery. Words only detract from the visceral resonance of the moment. But as we embark upon our quest to discover shadows of the incomprehensible, it's important to avoid any confusion. We are not in search of mysteries—of anything that brings an end to rational discussion. Our goal is to see if our very rationality leads us to inexplicable realities—paradoxes—or, at the very least, realities so counter to intuition they leave us dumbfounded. We're then left to ask if these inexplicable realities are telling us something. Are they shadows of things beyond comprehension?

References and Further Reading

1. Locke, John. *An Essay Concerning Human Understanding, Volume 1, Books 1 and 2.* 1690. Project Gutenberg. www.gutenberg.org/ebooks/10615. Accessed July 11, 2016.

2. ———. *An Essay Concerning Human Understanding, Volume 2, Books 3 and 4.* 1690. Project Gutenberg. www.gutenberg.org/ebooks/10616. Accessed July 11, 2016.

3. McGinn, Colin. "The Problem of Philosophy." *Philosophical Studies* 76, no. 2–3 (1994): 133–56.

4. Nagel, Thomas. "What Is It Like to Be a Bat?" In *Mortal Questions*. Cambridge: Cambridge University Press, 1979. First published 1974.

5. Searle, John. *The Rediscovery of the Mind*. Cambridge, MA: MIT Press, 1992.

I dream my painting and then I paint my dream.

—Vincent van Gogh

CHAPTER TWO
Torricelli's Trumpet and Koch's Snowflake

Evangelista Torricelli was a great admirer of Galileo Galilei and had the good fortune to meet Galileo shortly before the luminary died. By that time, the younger Torricelli was already developing his own reputation as a keen mind. He would go on to make numerous contributions in science and math, and is frequently cited as inventor of the barometer. But one discovery in particular helped establish Torricelli's reputation.

In 1641, Torricelli produced a container that confounded mathematicians and philosophers alike. Its shape, shown in Figure 2-1, is that of a horn, which is why it's sometimes called Torricelli's trumpet. What's special about Torricelli's trumpet is that it doesn't end in a mouthpiece. Instead, the horn gets ever smaller but never actually comes to a point. At 1 on the measuring stick, the diameter of the horn is a foot; at 2, the diameter is half a foot; at 3, a third of a foot, and so on. Reoriented with the bell facing upward like a container, the horn has no bottom. A dimensionless water drop—something mathematicians like to work with—would continue down the trumpet forever.

Using mathematical techniques that were quite novel at the time, Torricelli was able to show that the figure had a very surprising charac-

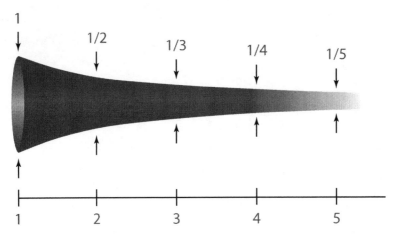

Figure 2-1: Torricelli's trumpet

teristic: finite volume. It holds a little less than six gallons. Try to pour any more in, and it will spill over the top.

Suppose we attempt to fill the container one drop at a time. Where would the first drop come to rest? With no bottom, it would continue down forever. So what would the other drops rest on? Wouldn't they all continue falling forever? And if so, how could the container ever be filled?

Torricelli's discovery caused quite a stir in its time. Mathematicians of the day were only beginning to grapple with infinity, and the horn and its ramifications captured their imagination. Torricelli's friend and mentor—a leading mathematician of the era—responded to a letter from Torricelli, writing, "I received your letter while in bed with fever and gout...but in spite of my illness I enjoyed the savory fruits of your mind...And having spoken of [your discovery] to some of my philosophy students, they agreed that it seemed truly marvelous and extraordinary."[1]

Torricelli recognized the significance of his work and passed along information about his discovery to Father Nicéron in Paris; from there

---

[1] Paulo Mancosu and Ezio Vailati, "Torricelli's Infinitely Long Solid and Its Philosophical Reception in the Seventeenth Century," *ISIS* 82, no. 1 (1991): 51.

it was widely disseminated to scholars throughout France, including such mathematical greats as Marin Mersenne, Gilles de Roberval, and Pierre de Fermat. Word of Torricelli's celebrity made its way back to Italy through Cardinal Ricci, who informed Torricelli, "Father Nicéron writes me that all the excellent men of that kingdom desire to see your works and that the sheet of propositions I sent them is passing through everyone's hands with great praise of your beautiful discoveries."[2] Torricelli was even offered Galileo's former position at the University of Pisa by a member of the Medici family.

But many leading philosophers weren't so thrilled. How could mathematics be trusted if it led to such intuitively ridiculous results? Empiricists like John Locke and Thomas Hobbes believed all knowledge came from what we experience in life. Since we can't experience infinity, they argued, we can't make definitive statements like "Torricelli's infinite horn has finite volume." To connect the finite, which we can experience, with the infinite, which we can't, was folly. Philosophers from all walks found Torricelli's result an enigma. Even Torricelli appreciated the broad repercussions of his work, referring to his discovery as "paradoxical."

Today we don't find the result quite so paradoxical, but we should. Mathematicians so routinely work with infinity that infinite figures like Torricelli's trumpet are second nature. Experience tells us that if we attempt to fill a bottomless container with paint, it will simply continue to flow downward. Yet if we take a mere six gallons of paint, we can fill Torricelli's trumpet to the brim and have paint left over. How can this be?

Torricelli's trumpet still remains an object of curiosity but for reasons Torricelli and his contemporaries were unaware of at the time. It turns out that the figure not only has finite volume but also has infinite surface area. And as math teachers like to point out, while the trumpet can be filled with paint, the prospect of painting it seems challenging

2 Ibid.

to say the least. How can a container with finite volume possibly have a surface that's infinite?

It's worth considering the possibility that mathematicians have come up with strange definitions of "volume" and "surface area" for figures that extend to infinity. For example, it's easy to agree that the box in Figure 2-2 has a volume of one cubic foot. We learn to calculate the volume of a box using the formula height × width × depth. We can't apply that formula to Torricelli's figure both because it has curves and because it extends to infinity. Mathematicians must therefore define volume in purely mathematical terms. The same is true of surface area. So perhaps it's the definitions themselves that are the source of trouble when applied to something like Torricelli's trumpet.

While the potential for muddled definitions is conceivable, it isn't the case. An important aspect of the development of calculus was for the purpose of calculating volumes and areas of arbitrary figures. When used to calculate the volume of a box or a ball, calculus gives a correct answer. In our finite world, everything about the definitions is consistent, correct, and based on a solid logical foundation. We're so

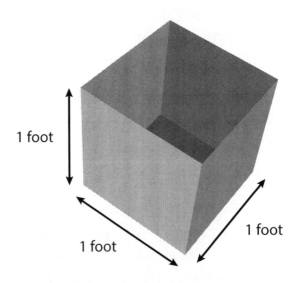

Figure 2-2: A one cubic foot box

certain of the definitions that when we apply them to infinite figures, we feel confident we can accept the results as true. Torricelli's trumpet has finite volume but infinite surface area because the definitions are sound—definitions that mathematicians, engineers, and scientists rely on with complete confidence. Which leads us once again to ask, how can Torricelli's trumpet have finite volume but infinite surface area?

Torricelli's trumpet isn't alone. Koch's snowflake, introduced by Helge von Koch in a 1904 paper, is an example of such a figure that can be understood using plane geometry. It also has the advantage that it doesn't run off to infinity in the same way as Torricelli's trumpet, though it does rely on infinity in a somewhat different fashion.

Koch's snowflake is constructed by looking at a sequence of ever more intricate shapes, each of which represents a better approximation of the Koch snowflake. The first is an equilateral triangle with all sides of length one foot. The second is formed by dividing each side of the first shape into three equal segments, constructing a new, smaller equilateral triangle with the middle segment as the base, then replacing the middle segment with the two sides of the newly constructed equilateral triangle. The construction is shown in Figure 2-3.

Note that the perimeter of the first shape is three feet in length. The second shape, however, has a longer perimeter. Looking at just a single one-foot-long side of the first shape, we see that the length of the

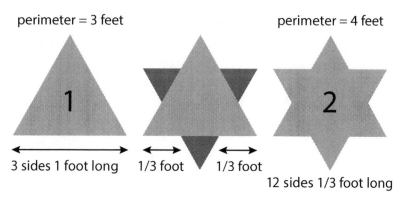

Figure 2-3: First step in approximating Koch's snowflake

corresponding part of the second shape (now consisting of four sides) is 4/3 of a foot. Since this is true for every side of the first shape, the perimeter of the second shape is 4/3 longer than the first shape.

$$\text{shape 2 perimeter} = 4/3 \times \text{shape 1 perimeter}$$

The third shape in the sequence is constructed in the same way. Each side of the second shape has its middle replaced with two sides of an equilateral triangle as shown in Figure 2-4. Since each side of the second shape is replaced by something 4/3 as long, the overall perimeter of the third shape is 4/3 as long as the second and $(4/3)^2$ as long as the first.

$$\text{shape 3 perimeter} =$$
$$4/3 \times \text{shape 2 perimeter} = (4/3)^2 \times \text{shape 1 perimeter}$$

In general, the nth figure constructed in this way has perimeter

$$\text{shape } n \text{ perimeter} = (4/3)^n \times \text{shape 1 perimeter} = (4/3)^n \times 3$$

The important point is that each figure has an increasingly longer perimeter, and the perimeter can be made as long as we desire. If we wanted a perimeter of at least one hundred feet, every figure starting with the thirteenth has this property. To reach a million feet, we'd have to get to the forty-fifth. We can't actually draw the figure

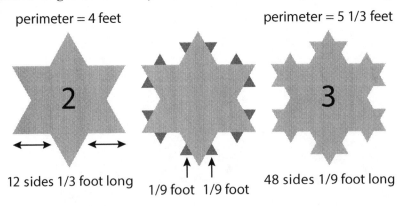

perimeter = 4 feet

perimeter = 5 1/3 feet

12 sides 1/3 foot long

1/9 foot   1/9 foot

48 sides 1/9 foot long

Figure 2-4: Second step in approximating Koch's snowflake

corresponding to $n = \infty$ anymore than we can actually draw Torricelli's trumpet. But we can we can talk about Koch's snowflake—the figure we ever more closely approximate—by considering larger and larger values of n.

By the way the successive figures are constructed, the size of the interior grows at each step just as the size of the perimeter does. Unlike the perimeter, however, the interior doesn't grow toward infinity. No matter how large n gets, a geometric argument can be made that each figure is contained in the circle defined by the corners of the original triangle (see Figure 2-5), so none of the sequentially constructed figures can have an area bigger than this circle. In point of fact, none of the sequential figures has an area that's even 8/5 as large as the original triangle, or about 7/10 of a square foot.

So far we've limited ourselves to the two-dimensional plane. If we raise the original triangular shape out of the plane by a foot, we get something that looks like a very large cookie cutter as shown in Figure 2-6. If we were to do the same with Koch's snowflake and set it on a flat table, creating what we might call Koch's cookie cutter, we'd be able to fill it with paint because the volume is finite. But painting it would prove far more challenging. Like Torricelli's trumpet, Koch's cookie cutter has infinite surface area.

Of course, we can't actually construct Koch's cookie cutter since it has an infinite number of sides. We can, however, construct an approxima-

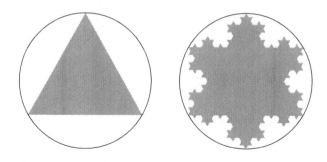

Figure 2-5: Containment circle for Koch's snowflake

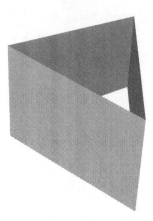

Figure 2-6: First approximating shape of Koch's cookie cutter

tion corresponding to any particular value of n. This in turn allows us to perform the following paradoxical thought experiment.

Imagine you're given a cubic foot of paint (about 7 1/2 gallons), and you're asked to paint a sheet of metal 1 foot wide and 100 football fields in length—a total of 30,000 square feet. You realize you can never get the paint to stretch that far. So you decide to fold the strip into an approximation of Koch's cookie cutter.

When n is small, the resultant cookie cutters are quite large. The value $n = 1$ yields a raised triangle 10,000 feet on a side. But as n gets larger, the associated cookie cutters shrink—the result of more folds with a fixed-length sheet of metal. A little math shows that the cookie cutter corresponding to $n = 32$ has a volume a bit less than one cubic foot, so you can set it on a table and fill it with the paint you have.[3] Do so, then lift it up for the excess paint to drain. The paint remaining stuck to the surface has done the job. The sheet of metal is painted.

---

[3]Admittedly, this would take a lot of folding: roughly 13 billion, billion folds.

References and Further Reading

1. Mancosu, Paulo, and Ezio Vailati. "Torricelli's Infinitely Long Solid and Its Philosophical Reception in the Seventeenth Century." *ISIS* 82, no. 1 (1991): 50–70.

Thus far I have explained the phenomena of the heavens and of our sea by the force of gravity, but I have not yet assigned a cause to gravity.

—Sir Isaac Newton

CHAPTER THREE
Gravity

Here's a magic trick you can easily perform for friends. Take a favorite stuffed animal and place it at one end of your kitchen table. Next, take a seat across the table and out of reach of the animal. Now proclaim to everyone present that you will cause the animal to slide across the table and into your hands. Stare intently at it. Stare some more. Build the suspense. Finally, after a good ten seconds, look your friends in the eye and proclaim, "What? Are you crazy? I can't magically make it move without touching it." To make your point, walk over to the animal and carry it back to your seat.

Now you're ready for the payoff. With everyone nodding in agreement, raise the animal in front of you and let it go. Without any visible means of propulsion, the animal drops from your hands to the floor. Gravity, it seems, is a magical force.

When first presented with this trick, most people are unimpressed. We're so familiar with gravity that it doesn't seem especially magical to see something drop to the floor. We'd be surprised if we let something go and it didn't drop. Yet we shouldn't allow our familiarity with gravity impede our appreciation for something scholars struggled with for centuries.

One of the great turning points in all of history arrived with Isaac Newton's theory of gravity, published in his masterwork the *Principia* in 1687. Decades earlier, Johannes Kepler carried out laborious calculations that led him to discover remarkable geometric facts about the movement of the planets. Remarkable as they were, however, they were simply facts, devoid of any underlying scientific theory to explain them. Kepler's discoveries proved fodder for many scholarly inquiries, but it was Newton who ultimately put the pieces together in a mathematically rigorous way. Newton not only developed a theory of gravity but went on to develop a general theory of motion that we now refer to as classical, or Newtonian, mechanics. Gravity became a special case of Newton's theory of motion, but one with a special twist.

At the foundation of classical mechanics are Newton's three laws of motion, the first of which is the following:

> *Newton's First Law of Motion:* An object at rest will remain at rest unless acted upon by an external force. An object traveling in a straight line at a constant speed will continue to do so unless acted upon by an external force.[1]

Examples are easy to cite. A kitchen chair will remain motionless until someone or something moves it. A billiard ball will continue across a billiard table in a straight line and at constant velocity until it encounters a cushion or another ball. Notice that here, as in most examples that jump to mind, something physical actually causes the

---

[1] The second assertion, that an object traveling in a straight line at a constant speed will continue to do so unless acted upon by an external force, represented a significant step forward in our understanding of motion. Day-to-day experience tells us the natural state of an object is stationary. If I give a box a shove along the floor and then let go, it quickly comes to rest. Newton's first law tells us the opposite: the box should keep moving unless acted on by an external force. We now recognize this force as friction, but for most of human history, it was assumed that force was required to keep an object moving because the natural state of an object was at rest. Even today children grow up making this assumption since it works in practice.

movement. I may pick up and move the chair, or it may be knocked over by a carelessly wielded vacuum cleaner, or it may be pushed across the floor by a strong gust of wind. It's natural to equate force with something physical—me, a vacuum cleaner, or molecules of air. And it seems mystical to imagine applying force to an object without something actually touching it.

Yet when it came to gravity, Newton appeared to be postulating just such a mystical entity. Here was the twist: the external force that kept the planets orbiting the sun rather than traveling off in a straight line was...what?

The idea of such a mystical force was met with incredulity by many. Christiaan Huygens dismissed the underpinnings of Newton's work as "absurd."[2] Gottfried Wilhelm Leibniz called Newton's gravity a "return to occult quantities, and even worse, to inexplicable ones."[3] Scholars— natural philosophers—were seeking to understand nature on its own terms, not through deference to something as unnatural as Newton was proposing.

Complicating the situation was the rise of what came to be known as mechanical philosophy. For centuries natural philosophy was dominated by the theory of substantial forms. The theory traces its roots to Aristotle, though it was refined and embraced by Scholastics beginning in the twelfth century. The theory was convoluted, expansive, and differed greatly in detail from one articulation to another. But a core precept was that an entity's substantial form was that which made it what it was—its essence. Within this context it would be perfectly acceptable to claim that a physical object, when dropped, fell to earth simply because that was its nature.

By the time of Newton, the theory of substantial forms was not only discredited in many circles but viewed with disdain. Descartes

[2] Brian Clegg, *Gravity: How the Weakest Force in the Universe Shaped Our Lives* (New York: St. Martin's Press, 2012), 73.
[3] Ibid.

dismissed it as "a philosophical being unknown to me,"[4] while Henry Oldenburg expressed his congratulations to Robert Boyle for having "driven out that drivel of substantial forms" that had "stopped the progress of true philosophy."[5]

Mechanical philosophy sought to do away with the ethereal nature of a thing's essence and instead focus on substance. The universe was to be conceived as consisting of matter in motion, and everything from sight to sound to the movement of chairs could be described in mechanistic terms—in terms of physical objects, large and small, colliding with one another. Sight was the result of light striking the eyes; sound the compression of eardrums by waves propagated through air; and chair movement the consequence of another object coming in contact with it. Adherents of mechanical philosophy were on the right track. But the theory of substantial forms was so deeply embedded in the reigning Scholasticism of the day that mechanical philosophers were quick to jump on anything that couldn't be traced back to a physical force—to objects colliding with other objects.

Set against this backdrop, it's not difficult to see why Newton's theory of gravity came under scrutiny. Not only did gravity seem outlandish on the surface (objects moving one another without physical contact) but it also smelled of the theory of substantial forms—a theory the new generation of scholars were seeking to exorcise from the world of natural philosophy.

Still, Newton's theory was so compelling that people sought to reconcile it with mechanical philosophy by providing various mechanistic means by which gravity might work. The common theme was to introduce some physical substance filling empty space— "ether"—thus providing a medium through which gravity could act. If gravity was transmitted from one object to another by way of contact with the ether, Newton's gravity and mechanical philosophy could peacefully coexist.

---

[4] Robert Pasnau, *Metaphysical Themes, 1274-1671* (New York: Oxford University Press, 2011), 549.
[5] Ibid.

Newton's views changed over time. He was initially an advocate of ether, but by the time of the writing of the *Principia*, he had shifted to a noncommittal position, leaning away from ether and toward action-at-a-distance without an intervening medium. These leanings were what evoked the contempt of individuals like Huygens and Leibniz. What Newton had come to fully embrace was a key tenet of modern science: science can only address propositions that can be experimentally verified. Until the existence of ether could be substantiated, Newton was satisfied to view gravity for what it was—a theory that correctly explained how the planets and other bodies moved. There was no need to say anything more. In the second edition of the *Principia*, published some twenty-six years after the original, Newton included an appendix known as the *General Scholium* in which he wrote:

> I have not as yet been able to discover the reason for these properties of gravity from phenomena, and I do not feign hypotheses. For whatever is not deduced from the phenomena must be called a hypothesis; and hypotheses, whether metaphysical or physical, or based on occult qualities, or mechanical, have no place in experimental philosophy.[6]

While the development of ether theories continued for centuries, scientific advancement didn't hinge on their success. Just as Newton had sidestepped ether by focusing on the fact that his theory worked rather than on the cause of gravity, physicists contented themselves with the notion of a field. A field is simply a convenient mathematical description of the gravity an object exerts at any point in space without any mention of its underlying cause. It's a mathematical way of saying gravity just *is*.

Which leaves us to again ponder this mysterious force that brings things crashing to the floor. Gravity reaches out and pulls on every-

---

[6] Isaac Newton, *Philosophiae Naturalis Principia Mathematica,* trans. by I. Bernard Cohen, Anne Whitman, and Julia Budenz (Oakland, CA: University of California Press, 1999), 943.

thing, yet we can't point to anything that's doing the pulling. Is gravity beyond comprehension?

It's a question we'll return to in a moment. However, it's first necessary to take an important detour to look at a discovery so contrary to how we perceive the universe that it left physicists dumbstruck.

*****

Not surprisingly, ether is an ethereal term. It's been proposed in many different contexts for many different reasons, and its actual nature is related to what it seeks to explain. By the early eighteenth century, the search for ether had shifted from finding a motive substance for gravity to finding a medium for the transmission of light. Thanks to the efforts of a myriad of people, and culminating in the work of James Clerk Maxwell and what came to be known as Maxwell's equations, the theory of light as a wave had gained acceptance. But if light was a wave, this in turn appeared to imply the need for a medium through which it could propagate.

The need for a medium of propagation comes from our experience with waves. The very nature of a wave cries out for a substance to carry it. What would it mean to speak of a wave rolling in off the ocean if it wasn't for the water? Sound provides yet another example. We hear sound because it's propagated as waves through the air. Strike a hammer on a laboratory desk and the sound startles us. Take away the air—create a vacuum—and the same action yields no sound at all.

Still, while it seemed a space-permeating ether should exist, it remained to be experimentally verified. Enter Albert Michelson and Edward Morley, who made use of the following observation in their search for ether.

Imagine a river flowing at a rate of 3 feet-per-second. Two swimmers set out from the same diving platform near one bank of the river as depicted in Figure 3-1. One swimmer swims the 100 feet directly across the river and touches a smaller platform before returning to the starting point. The other swimmer stays next to the bank, swimming upstream 100 feet before touching a platform and returning to the starting point. Both swimmers swim at an identical 5 feet-per-second in motionless water. Who returns to the starting point first?

Our immediate thought is that they return at the same time. Both have traveled 200 feet at the same speed. But have they? The swimmer who travels upstream and downstream fights the current for the first leg of the trip and then benefits from the current during the second. His net speed is 2 feet-per-second going upstream and 8 feet-per-second going down. The swimmer who swims across the stream fights the current, too, but in a more subtle way. If he didn't, if he put all his energy into swimming perpendicular to the current, he'd find that when he returned

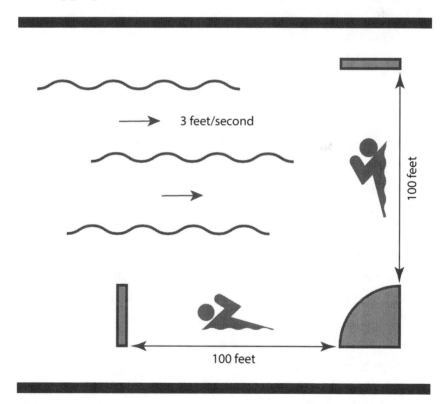

Figure 3-1: Two swimmers, one swimming up and downstream, one swimming cross stream

to the bank from which he began, he'd be downriver from the starting platform. But because he's swum directly across the river and returned to the starting point, part of his energy's gone into fighting the current. He achieves this by turning slightly into the river's flow—just enough so that he sees he's making his way directly across the river.

We might guess that the effects of the river cancel out, but they don't. When we work out the math, the cross-stream swimmer's trip is shorter than that of the upstream and downstream swimmer: 50 to 62.5 seconds. We can, in fact, calculate the time it would take for a swimmer to swim a distance of 100 feet in any direction and return to the starting point, as depicted in Figure 3-2. The key point is that the length of the trip is different depending the swimmer's angle relative to the flow of the river.

Michelson and Morley realized how to use this observation to develop an experiment to validate ether's existence. If ether were an invisible medium that permeated all of space, the earth and every other celestial body would be moving through it.

To make the swimming analogy work, rather than thinking of the earth moving through the ether, we can equivalently think of the earth standing still in the midst of a "river" of moving ether. (The math works the same way, just from a different perspective.) Now, instead of sending swimmers off in different directions to cover a 100-foot length and back, we'll send light waves. Since ether was hypothesized to be the medium that allowed light to propagate, our knowledge of waves leads us to conclude that the light waves should be affected by the ether in the same way that the swimmers are affected by the river of water. In particular, if we send two light waves off in different directions, we should find they return to their starting point at different times.

Of importance is that it's not necessary to know which way the river of ether is flowing (equivalently, what direction the earth is moving through the ether). The mere existence of a time difference between two waves of light would serve as evidence for ether.

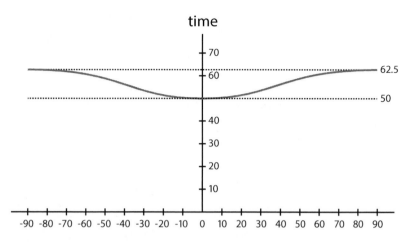

Figure 3-2: The time in seconds required to swim round trip as a function of the angle with the current; 0 degrees is perpendicular to the river flow, -90 degrees is upstream then downstream, 90 degrees is downstream then upstream

Michelson and Morley built a device that allowed them to send two light waves at perpendicular angles over identical distances so they could observe the difference in travel time by the waves. The device was placed on a rotating slab so it could be positioned at different angles relative to the ether.

To their surprise, no matter how they positioned the device, the two light waves always returned to their starting point at exactly the same time. Experimental fact was at odds with what was known about waves. Something was askew. If ether existed, Michelson and Morley should have observed different travel times by the two waves of light. But if the ether didn't exist, by what means could light waves propagate? The apparent contradiction was cause for concern. But it also provided food for thought.

If ether existed, it provided a universal backdrop against which things moved. Returning to our water analogy, imagine a boat floating in the middle of a calm ocean far from land. An observer on the boat can determine her speed by measuring how fast the boat is moving relative to the water. She looks over the side of the boat, watches the water against her oars, and sees how fast the water is passing by. The situation's the same for an astronaut traveling through deep space *if* there exists ether to play the role of the water. If there isn't, if space is void, what is she measuring her speed against?

In fact, how does she know if she's moving at all? Imagine riding in an airplane with no windows. The air is smooth and free of turbulence. Are you traveling at 400 miles per hour relative to the ground or 200? Or are you parked motionless at the gate? If the plane's traveling at a constant velocity—including standing still—the laws of physics work the same. If you drop a ball, for example, it falls straight to the floor no matter how fast the plane's moving. Galileo made this observation as long ago as 1632, though he used the inside of a sailing vessel rather than a plane to make his point.

Now imagine two astronauts, Alex and Anna, passing by one another at constant velocities in deep space. Alex sees Anna moving at, say, 100 feet per second, and Anna sees Alex moving in the opposite direction at the same speed. The question is who's moving, and who's standing still? Without ether, there's no fixed reference frame against which to measure the astronauts' speeds. If we assume that Alex is at rest and Anna is moving, it completely fits the facts. If we assume the opposite, that Anna's at rest and Alex is moving, it also fits the facts.

One point of clarification needs to be made with respect to what we mean by "moving." If either Alex or Anna is accelerating, all bets are off. We know from experience that when a pilot revs the engines and sends a plane racing down the runway, we feel ourselves pushed into our seats, and a dropped ball moves a bit toward the back of the plane as it falls to the floor. If either Alex or Anna is accelerating, they could feel motion. It's only when they aren't accelerating that we can arbitrarily choose who's at rest and who's moving.

A plane or a spacecraft or anything else that isn't undergoing acceleration is said to be in an *inertial frame of reference*. We can therefore summarize our discussion a bit more precisely, as follows.

With no ether to provide a universal backdrop against which things move, we can arbitrarily choose which of two inertial reference frames is at rest and which is moving.

Nothing we've discussed so far is remotely controversial when it comes to activities that involve physical motion, like the bouncing of balls. It therefore seems natural to expect all the laws of physics to behave the same in different inertial reference frames. Unfortunately, Maxwell's equations describing light waves were different when translating from one inertial reference frame to another.

This was a source of frustration for physicists, who expected more elegance from the universe. But it also served as motivation to provide an explanation. And that explanation came from Albert Einstein and his famous *special theory of relativity*.

In the first of the theory's two postulates, Einstein began by reaffirming physicists' expectations.

> *The Principle of Relativity:* The laws of physics are the same in all inertial frames of reference.

He then made a radical proposal in his second postulate.

> *The Constancy of the Speed of Light in Vacuum:* The speed of light in vacuum has the same value c in all inertial frames of reference.

To understand the fantastic nature of this postulate, let's first look at a simple experiment: throwing tennis balls. Imagine we set a tennis-ball launcher on the ground 100 feet from a target and observe that balls launched from the machine reach the target in exactly 1 second. From this we conclude that the launcher expels balls at a speed of 100 feet-per-second.

Consider performing the same experiment but this time with the launching machine and target strapped to the floor of a boxcar traveling at a constant speed of 10 feet per second on a flat, straight section of train track. Since the equipment and the observer are all moving at the same constant speed, the experiment will look identical to the experiment performed on the ground. Inside the boxcar, the ball will travel the 100 feet from the launching machine to the target in 1 second. An observer in the boxcar would conclude that the balls are being launched at 100 feet per second, which, in fact, they are relative to the frame of reference provided by the boxcar.

Now perform the same experiment after removing the walls and ceiling of the boxcar, as in Figure 3-3, and consider what an observer standing on the ground would see. She would see the ball leave the launcher and arrive at the target one second later, exactly as the observer in the boxcar did. But she would also see that the ball had traveled 110 feet during that time relative to the ground. The ball traveled the extra distance because the boxcar, and therefore the target, was moving at 10 feet per second during the time the ball was in flight.

There's nothing counterintuitive about this observation. The fact that the speed of the ball relative to the ground is the combined speed of the train and the launcher makes complete sense. The result isn't limited to hard objects like balls. Sound waves, for example, perform in the same fashion, which is why a police siren sounds different when it's coming toward you than when it's moving away.

Returning to Einstein's second postulate, assume that instead of launching tennis balls in our experiments, we send out a beam of light. According to the postulate, the speed of light as measured by an observer on the ground isn't any faster than it was on the train. Somehow, light doesn't get a boost by the train's movement. This is true if the train is moving at 10 feet per second or 100 million feet per second or any other speed.

At an intuitive level, this contradicts all common sense. If we observed the same behavior with tennis balls, we'd be in disbelief. But following the mathematical trail blazed by Einstein's postulates leads to several positive results. First, Maxwell's equations proved to work

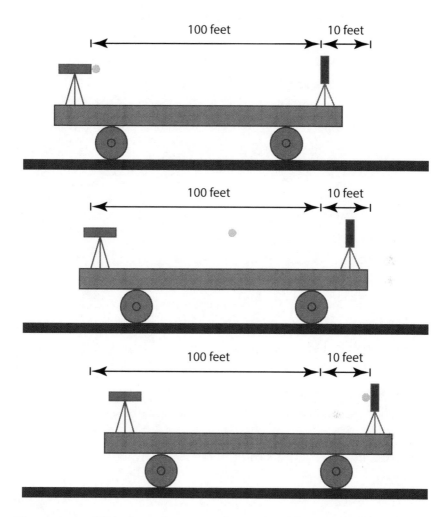

Figure 3-3: With the train car moving to the right at 10 feet-per-second,
the ball travels 110 feet in one second relative to the ground

the same in all inertial frames of reference—most of Einstein's paper on special relativity was devoted to showing just this fact. In addition, special relativity was both simple and simplified how the laws of physics fit together. With all inertial reference frames placed firmly on an equal footing, the idea of one special reference frame provided by ether was seriously called into question. And the theory was in agreement with Michelson and Morley's experimental results.

Yet following the mathematical trail of Einstein's postulates also led to equally surprising consequences—consequences seemingly more counterintuitive than the second postulate's claim that the speed of light was always the same for different observers.

Imagine once again our two astronauts, Alex and Anna, floating weightlessly together in a spaceship. Each has a watch, and once they set their watches, each watch clicks forward in unison. When Alex sees nine o'clock on his watch, so does Anna. When Alex sees ten o'clock, Anna does, too.

Now we change the scenario. The watches remain just as they are, but we put Anna in a different spacecraft that's headed toward Alex at close to the speed of light. Both float weightlessly in their respective spaceships as neither one is accelerating. At the moment the two spacecraft pass one another, Alex looks out his window at Anna. What does he see?

The first thing Alex might notice is that Anna's watch is ticking more slowly than his. Anna actually looks like a movie running in slow motion. Another thing Alex would notice is that Anna, along with everything in her ship, is visually distorted—pressed flatter in the direction she's moving. Even more astonishingly, if Anna looked out her window, she'd see exactly what Alex sees. To Anna, Alex's clock is ticking more slowly than hers, and Alex looks flatter.

These examples only hint at the bizarre implications of special relativity. Another is the twin paradox. Suppose Alex and Anna were twins living on earth. Anna leaves on a spacecraft that travels to the nearest star at near the speed of light. Upon her return, she finds that while she has aged a mere two days, Alex has aged many years. Anna is still young and full of life, but Alex is an old man with gray hair. The difference in their ages is determined by the speed Anna was traveling. The closer she traveled to the speed of light, the greater the difference.

It would seem that the inherent strangeness of special relativity is beyond comprehension. However, if we accept Einstein's postulates, then everything else falls out. Every prediction made by special relativity, no matter how seemingly preposterous, can be explained in terms well within our capability to conceive of them.

We come to believe that time and space are immutable because in our daily lives that's how we experience them. Time clicks forward at the same rate for me and for the jogger running past me. The tree outside my window looks the same whether I'm sitting in my office or driving by in a car. We develop these conceptions because we move so slowly relative to the speed of light. And they seem so clear, so obvious, that it's difficult to overcome them. The difficulty for students learning the special theory of relativity isn't the math, which is quite simple. The real difficulty is breaking free of closely held preconceptions, which, in turn, cause students to struggle through the learning process.

At the heart of special relativity, however, is the paradoxical observation made by Michelson and Morley, later codified in Einstein's second postulate: *the speed of light in vacuum has the same value c in all inertial frames of reference.* Light doesn't behave like tennis balls or sound waves or anything else we know of. It doesn't behave as our minds tell us it must, or at least, as it should.

What Einstein did with his special theory of relativity was to inextricably link time and space. Prior to his efforts, these two most basic of concepts were viewed as separate. Time seemingly ticked away at the same pace everywhere in the universe. Space, or rather our movement through it, had no effect on the answer. Einstein made it clear that this isn't correct.

Hermann Minkowski, a mathematician and one of Einstein's teachers, showed how Einstein's ideas could be expressed using a simple mathematical framework that interlocks the time variable t with the three space variables x, y, and z. It's this framework that gave rise to the name *spacetime*. Einstein initially dismissed Minkowski's math as "superfluous learnedness."[7] However, Einstein soon changed

---

[7] Abraham Pais, *Subtle is the Lord: The Science and Life of Albert Einstein* (Oxford: Oxford University Press, 1982), 152.

his mind, adopting and extending Minkowski's spacetime geometry in what was to become his masterpiece.

*****

As special relativity came to be accepted by the scientific community, one consequence was the disappearance of ether. It wasn't needed, the Michelson-Morley experiment provided evidence that it didn't exist, and physicists had grown comfortable that light, as expressed in Maxwell's equations, didn't need a propagating medium. But what about gravity? What about a substance that explained how gravity worked so that it wasn't a mystical action-at-a-distance?

The answer again came from Einstein, but this time from a new theory, the *general theory of relativity*. What makes special relativity special is that it considers only inertial reference frames—reference frames that experience no acceleration.

Einstein took his first step toward general relativity by connecting acceleration with gravity. To understand how, imagine an astronaut in a spacecraft with no windows. Now, rather than floating freely in space, we turn on the engines, causing him to accelerate forward at a uniform rate. What does the astronaut experience? In the same way that an accelerating airplane pushes you back in your seat, the astronaut will feel the push of the floor against his feet. Should he step on a scale, it would register weight. If he were to hold a ball in front of him and let go, the ball would fall to the floor.[8] Should he want to, the astronaut could set up a basketball court and play a perfectly good game of basketball. With the ship accelerating at just the right rate, the game would play exactly as if he were on earth (a little slower and he could easily jump and dunk the ball; a little faster and he'd have trouble getting the ball to the hoop). In fact, with no windows, how

---

[8] In actuality, once separated from the astronaut's hand, the ball will continue in the direction of the ship at a constant velocity while the floor of the ship, still accelerating, will catch up to the ball

does he know the ship isn't simply standing on the earth's surface? Instead of engines pushing his spacecraft forward, perhaps it's simply parked at a playground somewhere.

Just as Einstein hypothesized in special relativity that light moves at the same speed in all inertial reference frames, when developing the general theory, he assumed the laws of physics were the same for someone standing in a gravitational field and someone moving in a uniformly accelerating reference frame. This *principle of equivalence,* as it came to be known, helped start Einstein on a multiyear quest. And just as with special relativity, some of the consequences of general relativity were surprising, such as the fact that clocks run more slowly in the presence of gravity.

Yet perhaps the most celebrated aspect of general relativity was its ability to describe gravity as something other than an inexplicable force. As Einstein once again set about to describe the universe through equations linking time and the three space variables, he found the task considerably more difficult than the development of special relativity. It wasn't until ten years after the publication of the special theory that a paper appeared describing the general theory and *Einstein's field equations.*

One characteristic of Einstein's field equations is that they weren't the simple linear equations found in Minkowski's description of special relativity. If Minkowski's spacetime equations are likened to a flat sheet of paper, then the spacetime equations of general relativity can be likened to that same sheet of paper after it's been warped by bending and twisting it (without folding). Just as the equations of special relativity describe how time and space are related between any two inertial reference frames free of the influence of gravity, the equations of general relativity describe how time and space are related when matter and gravity are incorporated. The conventional way of describing this state of affairs is to reject the notion of gravitational force as an actual force, and to say that stars, planets, and all other forms of matter simply warp spacetime.

It would appear, then, that the concerns dating back to the time of Newton—that gravity was a mysterious force—had been laid to rest. Gravity is just a warping of spacetime. But what exactly does this mean? It says gravity can be explained as a mathematical artifice. But we're still left to ponder. Is gravity beyond comprehension? As Newton's critics argued, when something falls to the floor without any visible motive force, should we be incredulous?

The question that's thrust front and center is whether or not the existence of a mathematical description, however elegant, is sufficient for us to declare that we now comprehend what's going on. Or if, instead, it's necessary to experience what the math actually means in the physical world. To stand up, go to the window, and declare that things now look different.

In a formal sense, we actually can experience the mathematical description. Countless experiments have been run that show spacetime is a physical reality. When the speed of light is measured by observers in different inertial reference frames, it's found to travel at the same speed. Two synchronized clocks—one left in the laboratory, one sent around the world on a plane—are no longer synchronized when the latter returns to its starting point. The difference in time is as predicted by relativity.

Yet, while we can measure the effects of general relativity, we don't actually see the warping of spacetime. For that matter, we don't "see" time. As humans, we see things in space, but we experience time cognitively. We're thus left to ask if the curvature of spacetime as captured in Einstein's field equations provides an intuitively comprehensible explanation of gravity.

Presumably, some would argue that it does. Individuals who spend their lives working closely with any set of equations develop insights about those equations through experience. And who's at liberty to say when the region between comprehensibility and incomprehensibility is entered, if not crossed? As for myself, I'll still occasionally toss a ball in the air and ask myself: what is it that makes it come down?

References and Further Reading

1. Clegg, Brian. *Gravity: How the Weakest Force in the Universe Shaped Our Lives.* New York: St. Martin's Press, 2012.

2. Collier, Peter. *A Most Incomprehensible Thing. Notes towards a Very Gentle Introduction to the Mathematics of Relativity.* Incomprehensible Books, 2014.

3. Gardner, Martin. *Relativity Simply Explained.* New York: Vintage Books, 1976.

4. Kochiras, Hylarie. "Force, Matter, and Metaphysics in Newton's Natural Philosophy." PhD diss., University of North Carolina, 2008. See also www.generativescience.org/papers/nature/Kochiras-_2008-240.pdf. Accessed July 11, 2017.

5. Manwell, Alfred R. *Mathematics Before Newton: An Inaugural Lecture Given in the University College of Rhodesia and Nyasaland.* New York: Oxford University Press, 1959.

6. Newton, Isaac. *Philosophiae Naturalis Principia Mathematica,* 3rd ed., 1726. Translated by I. Bernard Cohen, Anne Whitman, and Julia Budenz. Oakland, CA: University of California Press, 1999.

7. "Newton's Philosophy." In *Stanford Encyclopedia of Philosophy.* Last modified May 6, 2014. plato.stanford.edu/entries/newton-philosophy.

8. Pais, Abraham. *Subtle is the Lord: The Science and Life of Albert Einstein.* Oxford: Oxford University Press, 1982.

9. Pasnau, Robert. *Metaphysical Themes: 1274-1671.* New York: Oxford University Press, 2011.

In mathematics you don't understand things.
You just get used to them.

—John von Neumann

CHAPTER FOUR
A Closer Look at Infinity

As humans we may not have trouble with the number ten, but we nonetheless have our cognitive limitations when it comes to numbers. Relatively small numbers are easy for us. When I tell my spouse there are twelve eggs in the refrigerator but only three apples, there's no need to elaborate. She can picture twelve and three, and she has an immediate sense of how many more eggs there are than apples. Just as importantly, she can make meaningful use of this information relative to our lifestyle: we have plenty of eggs, but we may want to buy some apples next time we're at the store.

When we get to larger numbers, however, we start to have problems. A trillion is bigger than a billion. In fact, a trillion is a thousand billion. But that doesn't register in the same way as twelve is bigger than three. When we get to large numbers, we frequently rely on visual analogies, such as a picture of a billion dollars stacked next to a trillion dollars. Still, even though we don't conceptualize large numbers as easily or in the same way as smaller numbers, a billion or a trillion or a googol (10 to the power 100) isn't beyond our comprehension. We know what these numbers represent, and we know how to use them just like any other numbers when adding, subtracting, or performing other calculations.

Where we truly run into trouble is with infinity. We even run into trouble trying to define infinity. Is it a number that's larger than any other number? This defines infinity as an actual number, and is that what we want? Or we could say it's an unlimited quantity. That's better. But isn't unlimited just another word for infinite?

Definition or not, we do have a sense of infinity. If I begin counting the numbers one, two, three... I realize there's no point at which I must stop. Infinity is, in some sense, the end I'll never reach. We're comfortable with infinity even though it stretches our imagination to think about. Yet, as we think more carefully about infinity, it gives rise to some unsettling thoughts. Consider the following problem, a version of which was first proposed by Greek philosopher Zeno of Elea in the fifth century BCE. Suppose I'm standing in front of a wall. As I reach out to touch it, my hand must cross half the distance to the wall. It then must cross half the remaining distance, then half the remaining distance, and so on. Looked at in this way, my hand must traverse an infinite number of finite distances. How, therefore, can it actually reach the wall?

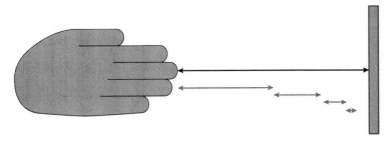

Figure 4-1: A hand reaching out to touch a wall

Most people run into some form of this paradox, known as the *dichotomy paradox*, at one time or another. Its durability is a testament to the fact that the problem is truly paradoxical. Mathematicians tend to disregard its paradoxical nature by relying on the notion of a *limit*. But limits don't address the fundamental issue at hand.

To understand the concept of a limit, it helps to look in somewhat more detail at the dichotomy paradox. If the distance from my hand to the wall is 1 unit when I begin to reach, it first travels 1/2 a unit, then 1/4, then 1/8, and so on. This in turn suggests that the infinite summation 1/2 + 1/4 + 1/8 + 1/16 +... is equal to 1.

Since we can't actually carry out such a summation, mathematicians instead focus on the following collection of partial summations, each consisting of one additional term.

1/2
1/2 + 1/4 = 3/4
1/2 + 1/4 + 1/8 = 7/8
1/2 + 1/4 + 1/8 + 1/16 = 15/16
...

Note that each partial summation gets ever closer to the value 1 and that no term ever exceeds 1. A formal way of saying this is that for any small number ε of our choosing, we can find a large enough number of terms n in the summation so that all partial sums with at least n terms are within ε of 1.

For example, suppose we chose $\varepsilon = 1/8$. Then, looking at our list of partial summations, we see that every term with $n \geq 3$ is within 1/8 of 1. The important point is that no matter what tiny value of ε we choose, we can find a corresponding value of n. In this case, mathematicians say that the infinite summation 1/2 + 1/4 + 1/8 + 1/16 +... approaches the value 1 in the limit as the number of terms n in the summation goes to infinity.

Nineteenth-century mathematicians slowly adopted this formal definition of a limit and, in doing so, helped place calculus and other fields of math dealing with infinitesimals on a solid foundation. Part of the beauty of this achievement, however, was that it avoided having to deal with infinity directly. The notion of a limit states that for any tiny but finite ε there exists a large but finite n for which a particular condition holds. When speaking precisely, no mathematician would say that $1/2 + 1/4 + 1/8 + 1/16 + ...$ equals 1. The individual would say that the infinite summation $1/2 + 1/4 + 1/8 + 1/16 + ...$ approaches 1 "in the limit as the number of terms in the partial summation gets ever larger."[1]

Recognizing this, the notion of a limit doesn't address the dichotomy paradox. In the example of a hand reaching toward a wall, a mathematician would say, "yes, in reaching toward the wall in the way described, the hand can get as close as desired by passing through a large, though finite, number of distances." As a mathematician, however, the individual would have to remain silent about actually reaching the wall. Clearly, however, the hand does, in fact, reach the wall. And as Zeno would point out, not only does the hand traverse an infinite number of distances, each individual distance has a length strictly greater than zero.

The dichotomy paradox isn't simply a matter of wordplay; it's a true paradox. How can we possibly traverse an infinite number of distances, each individually of length greater than zero, and reach the end of our journey? Yet it happens constantly. And our best mathematics not only fails to resolve the paradox, it further invites the question.

Aristotle, Hermann Weyl, Bertrand Russell—these are just a few of the historical names who weighed in on the metaphysical aspects of Zeno's many related paradoxes. Yet for the new insights provided by each contributor, the paradoxical nature of the problem persists.[2]

---

[1] In mathematical parlance, "in the limit as n goes to infinity," where the term "limit" is formally defined as we have done using the terms ε and n.

[2] Or it is replaced by something even more paradoxical. Various resolutions of Zeno's paradoxes have been put forward that rely on modern concepts of the nature of time and space. For example, if space is discrete in the sense that the hand can only occupy certain fixed positions along the path leading to the wall, then it wouldn't have to pass through an infinite number of finite distances, but only through a (large) finite number. But this in turn raises the paradoxical question of what it means for the hand to move from one discrete location to the next without passing through the space in between.

Infinite summations such as those found in the dichotomy paradox are a perpetual source of delight and consternation, and we needn't look too deeply to find them. We know from performing long division or simply typing 1 ÷ 3 into a calculator that

$$1/3 = .333333\cdots$$

where the sequence of 3s extends infinitely. Similarly, 2 ÷ 3 can be written as

$$2/3 = .666666\ldots$$

where again the sequence of 6s extends infinitely. We know that 1/3 + 2/3 equals 1. If, however, we add the two decimal expressions we get

$$.333333\cdots + .666666\ldots = .999999\ldots$$

thus yielding the apparent contradiction

$$1 = .999999\ldots$$

It seems that in performing the addition using the two decimal expressions, some very small amount gets lost in the shuffle. How can these two numbers be the same?

In actuality, what we're seeing is simply another example of the dichotomy paradox. Formally, the number .999999... is defined as

$$9/10 + 9/100 + 9/1000 + 9/10,000 +\ldots$$

But since this is the sum of an infinite number of terms, we speak instead of its limit of partial summations.

9/10
9/10 + 9/100 = 99/100
9/10 + 9/100 + 9/1000 = 999/1000
9/10 + 9/100 + 9/1000 + 9/10,000 = 9999/10,000
...

The nonterminating sequence of 9s simply says that these partial summations approach a value of 1 in the limit as we take more and more terms.

By convention, we say that .999999... = 1, and it doesn't get us into any trouble because summations associated with infinite decimals are well-behaved.[3] However, strange things can happen when we look at less well-behaved summations. In 1703 Guido Grandi looked at the following infinite summation.

$$1 - 1 + 1 - 1 + 1 - 1 + ...$$

Grandi then asked what it totaled by adding subsequent terms.

$$1 - 1 = 0$$
$$1 - 1 + 1 = 1$$
$$1 - 1 + 1 - 1 = 0$$
$$1 - 1 + 1 - 1 + 1 = 1$$

...

Calculated in this way, the sequence doesn't add up to any particular number. It simply continues to jump between 0 and 1. But suppose we perform the addition in a slightly different way.

$$
\begin{aligned}
& 1 - 1 \ + \ 1 - 1 \ + \ 1 - 1 \ + ... \\
= \ & (1 - 1) + (1 - 1) + (1 - 1) + ... \\
= \ & \quad 0 \ + \ \ 0 \ + \ \ 0 \ \ + ... \\
= \ & \quad 0
\end{aligned}
$$

Calculated in this fashion, the sum of all the numbers appears to be 0. But let's try the summation in yet another way.

$$
\begin{aligned}
& 1 \quad -1 + 1 \quad -1 + 1 \quad -1 + 1... \\
= \ & 1 + (-1 + 1) + (-1 + 1) + (-1 + 1) + ... \\
= \ & 1 + \quad 0 \ + \quad 0 \ + \quad 0 \ + ... \\
= \ & 1
\end{aligned}
$$

_____

[3] Formally, absolutely convergent.

From this calculation, the sum appears to equal 1. How can the same summation add up to both 0 and 1 at the same time?

Grandi wasn't finished. Since the total jumps back and forth between 0 and 1 when calculating the summation one term at a time, Grandi felt that the correct value of the infinite sum should be 1/2. And, in fact, he was able to show this. By applying long division to calculate $1/(1+x)$, he arrived at the expression

$$1/(1+x) = 1 - x + x^2 - x^3 + x^4 - x^5 + \ldots$$

Upon setting $x = 1$, Grandi found what he was looking for.

$$1/2 = 1 - 1 + 1 - 1 + 1 - 1 + \ldots$$

Grandi was a Jesuit, in addition to being a well-regarded mathematician. Upon demonstrating that the sum was both 0 and 1/2, he proclaimed that it showed God could create something from nothing. Mathematicians of the day weren't so moved, but Grandi's work helped serve as a catalyst to find out what was going on. Infinite sums were starting to pop up increasingly during Grandi's time, and they posed a threat to the orderly world of mathematics.

At the heart of Grandi's problem was that certain rules we associate with addition don't hold true when working with infinite series. No one needs to teach us that $(1 + 2) + 3 = 1 + (2 + 3)$. Nor do we need to be taught that $1 + 2 = 2 + 1$. In grade school we learn the names of these properties—associativity and commutativity, respectively—but every child knows that if a group of people throw all their pennies in a pile, it doesn't matter how they get there. You still end up with the same number of pennies. If the group happens to be infinite, however, you can't count on that being true. When summing together an infinite collection of numbers, we can get a different result depending on how we add them together. Adventurous readers are invited to take a look at the appendix to this chapter, which shows, through seemingly valid arithmetic manipulations, that $1 + 2 + 3 + 4 + \ldots = -1/12$.

As seen with Grandi's problem, it's not difficult to show that the order in which an infinite collection of numbers is added together can make a difference in the sum. Logic forces us to conclude that arithmetic operations that we know without being taught—operations that are perfectly valid when working with any finite collection of numbers—don't necessarily work with an infinite collection.

Mathematician David Hilbert treated us to yet another glimpse of how strange infinity can be with a clever example known as Hilbert's grand hotel. The hotel has an infinite number of rooms labeled 1, 2, 3... and so on. One day a visiting dignitary came looking for a room, but learned they were all occupied. The dignitary turned to leave, but was called back by the proprietor. "We may be full, but we still have room," said the proprietor, at which point he sent a note to all the guests asking them to move to the room that was numbered one higher than the room they were in. This in turn freed up room 1 for the dignitary.

Not only could the proprietor handle one extra guest, he could make room for an infinite number of new guests. How? By asking everyone in the hotel to move to the room with a number two times larger than his or her room number. This left the infinite number of odd rooms empty to house all the new guests.

Simple though Hilbert's example may be, it shows that the concepts of "full" and "no space available" aren't one and the same. For any real hotel, a hotel with a finite number of rooms, the distinction is pointless. A no-vacancy sign doesn't just mean every room is occupied; it means don't bother knocking at the door, since you're not going to get a room. At Hilbert's grand hotel, a no-vacancy sign might as well read "come on in." Once again, infinity defies some of our most deeply held convictions, leaving us unsettled. It's tough trying to peer over an edge that isn't there.

*****

Georg Cantor is best known for taking a radical, systematic look at infinity. His work differed so dramatically from anything before it that mathematicians weren't initially sure what to make of it. Some loved it; others hated it. David Hilbert believed Cantor had created a new "paradise for mathematicians."[4] Bertrand Russell viewed Cantor as one of the "greatest intellects of the nineteenth century."[5] Henri Poincaré, on the other hand, thought of Cantor's work as "a grave mathematical malady, a perverse pathological illness that would one day be cured."[6] Leopold Kronecker went even further, famously calling Cantor a "charlatan," a "renegade," and a "corrupter of youth."[7]

Cantor had shaken the mathematical world by realizing there were many types of infinity, some infinitely bigger than others. And his efforts weren't a parlor game. He helped propel the world of mathematics into a decades-long evaluation of its very foundations.

Cantor broke with mathematical tradition by focusing on objects that had received little, if any, attention: sets. He began with a simple question: how do we know when two sets are of the same size? When we consider the size of two finite sets, we can simply count how many items each set contains and compare the two numbers. The set of fingers on my right hand is of size 5, while the set of socks on my feet is of size 2, from which we conclude the sets are not of the same size. But this simple technique won't do for infinite sets.

We can, however, use a technique that doesn't require counting. If I look at a table filled with cups sitting on saucers, I can see if the set of cups and saucers have the same size without actually counting. I just have to confirm that each cup is sitting on exactly one saucer and every saucer has exactly one cup sitting on it. If so, we say the two sets are in one-to-one correspondence and are therefore of the same size.

[4] Joseph Warren Dauben, *Georg Cantor: His Mathematics and Philosophy of the Infinite* (Princeton, NJ: Princeton University Press, 1990), 1.
[5] Ibid.
[6] Ibid.
[7] Ibid.

If every cup sits on its own saucer but some saucers are without cups, the set of saucers is larger, and if every saucer has a cup but some cups are without a saucer, the set of saucers is smaller. These definitions of smaller, larger, and equal in size are not just reasonable, they seem obvious.

Consider now the set of all counting numbers {1, 2, 3...} and the set of all even numbers {2, 4, 6...}. Both sets are infinite, and our intuition tells us the first set is larger than the second since the second set is a strict subset of the first. We might even be tempted to say the set of counting numbers is twice as big as the set of even numbers.

However, under the definition we've adopted, that's not the case. As shown in Figure 4-2, we can find a one-to-one correspondence by associating the 1 in the first set with the 2 in the second set, the 2 in the first set with the 4 in the second set, and so forth. Stated another way, the operation "assign a number in the first set to a number in the second set by multiplying the first number by 2" describes a one-to-one correspondence between the two sets. Under our definition—the definition proposed by Cantor—the two sets are of the same size.

The result seems counterintuitive because of preconceptions formed by our experience with finite sets. "Counting" and "correspondence" are two ways of achieving the same result when working with finite sets. However, we don't have the luxury of being able to count infinite sets and so must turn to correspondence. One of the oddities of relying on correspondence is that, as we've seen, an infinite set and a strict subset of that set may have the same "size." To avoid this confusion, Cantor instead used the word *cardinality*: two sets have the same cardinality if they can be put in one-to-one correspondence. While cardinality is

$$\{1,\ 2,\ 3,\ 4,\ 5,\ \ 6,\ ...\}$$
$$\updownarrow\ \updownarrow\ \updownarrow\ \updownarrow\ \ \updownarrow\ \ \ \updownarrow$$
$$\{2,\ 4,\ 6,\ 8, 10,\ 12,\ ...\}$$

Figure 4-2: A one-to-one correspondence between two sets

intended to represent size at an intuitive level, it's formally about pairing things off in two different sets. Still, the fact that the "obviously" smaller set {2, 4, 6...} isn't smaller than {1, 2, 3...} when applying the one-to-one correspondence measure is unsettling and again points to the limits of our comprehension.

Infinite sets that can be put in one-to-one correspondence with the counting numbers are called *countable*. We might ask if all infinite sets are countable, in which case our definition leads us nowhere: infinite is infinite. But Cantor showed that's not the case by showing there's a set that isn't countable.

Consider, for example, all the real numbers on the unit line segment. For our purposes, we can think of these numbers as any decimal representation of a number between 0 and 1; numbers such as .75 and .276 are finite decimals. Also included are all decimal strings of infinite length, such as .3333... and infinite sequences that we might think of as "random." (Note that numbers like .75 can be extended to an infinite number of digits by adding an infinite number of 0s to the end.) Any number that can be expressed as a sequence of digits following the decimal point corresponds to a real number between 0 and 1, and any real number between 0 and 1 can be expressed with such a string of digits. We then ask, is the set of all real numbers between 0 and 1 countable? That is, can it be put in one-to-one correspondence with the counting numbers?

If so, it follows that we can put together a list of all such numbers. The number corresponding to the counting number 1 will be first in the list, the number corresponding to the counting number 2 will be second in the list, and so on, as shown in the hypothetical list depicted in Figure 4-3.

Cantor's now famous proof involved showing that whatever the list, he could construct a number that wasn't in it. This in turn allowed him to conclude that no such list exists and therefore that the numbers between 0 and 1 aren't countable.

He began by looking at the first number in our proposed list, .72463...
The first digit is a 7. The number we will construct, said Cantor, will
have as its first digit any digit other than 7. Arbitrarily pick 1.

Moving down the list, the second number is .39917..., which has
as its second digit 9. The second digit for the constructed number is
chosen as anything except 9. Arbitrarily pick 1.

Continuing, the third number is .44184..., which has as its third
digit 1. The third digit for the constructed number is chosen as anything
except 1. Arbitrarily pick 2.

$$
\begin{array}{c|l}
 & .\,1\ 1\ 2\ 1\ 2\,\ldots \\
\hline
1 & .\,⑦2\ 4\ 6\ 3\,\ldots \\
2 & .\,3\ ⑨9\ 1\ 7\,\ldots \\
3 & .\,4\ 4\ ①8\ 4\,\ldots \\
4 & .\,4\ 5\ 0\ ③9\,\ldots \\
5 & .\,0\ 2\ 8\ 7\ ⓪\,\ldots \\
\vdots & \qquad\vdots
\end{array}
$$

Figure 4-3: Cantor's diagonal argument that the set of
real numbers isn't countable

Notice that at every step the number from Cantor's construction
has the property that digit n doesn't match digit n in the nth number
in the list, as shown in Figure 4-3. Continuing the process forever, we
see that the number so constructed, .11212..., isn't in our proposed list.
We've used a specific list in Figure 4-3 for expository purposes, but our
construction would have worked no matter what the proposed initial
list. Therefore no such list exists.

Further, notice that at step n of our construction, we have nine separate digits we can choose (in the first step, we can choose any digit
other than 7). With nine choices, the number of numbers we could construct that are not in the list is 9 × 9 × 9 ×... In other words, whatever
our proposed list, there are infinitely more missing numbers than there

are on the list. Infinite sets that can't be put in one-to-one correspondence with the counting numbers are called *uncountable*.

Cantor had demonstrated something quite remarkable. Infinity wasn't just a single, elusive quantity. The counting numbers were infinite, but there were infinitely more real numbers found between 0 and 1. The result isn't in itself paradoxical, and, in fact, the distinction helped set mathematics on a new and better course. But it yet again helps us realize how elusive infinity is.

Cantor went even further, arguing the existence of an infinite number of infinitely larger infinities. The *power set* of a given set is the set of all subsets of that set. For example, the power set of the set $\{a, b, c\}$ is the set $\{\emptyset, \{a\}, \{b\}, \{c\}, \{a, b\}, \{a, c\}, \{b, c\}, \{a, b, c\}\}$, where $\emptyset$ is the empty set. For finite sets, it's easy to see that the power set must be larger than the original set. Cantor showed that for any infinite set, its power set had to be infinitely larger. Thus, by repeatedly creating the power set of any infinite set, it's possible to conclude there exist an infinite number of ever larger infinities.

Cantor tackled other questions of the infinite as well. One was the relative size of the unit line segment and the unit square, as depicted in Figure 4-4. Visually we can think of constructing the square by drawing all the line segments for fixed values of y. Figure 4-4 shows the different line segments corresponding to $y = .25, .5$, and $.75$. Of course, if we wanted to construct the entire square, we'd have to use all values of y between 0 and 1, and as we know from our earlier discussion, this set is not only infinite but uncountably infinite. So while the number of points in the unit line segment is uncountably infinite, the number of points in the unit square can be thought of as an uncountably infinite collection of sets each of which is uncountably infinite. Clearly, the set of points in the unit square must be larger than the set of points on the unit line segment. Or is it?

When Cantor first raised the question, he didn't get much of a response from the mathematical community. The answer seemed so

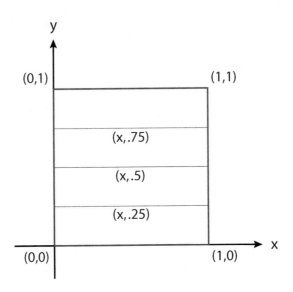

Figure 4-4: The unit square and several unit line segments

obvious that no one could get excited about it. Cantor's opinion, along with that of most other mathematicians who even bothered to consider the question, was that "the answer seems so clearly to be 'no' that proof appears almost unnecessary."[8] Still, Cantor was unable to find a proof.

Given his lack of success, Cantor changed course and attempted to prove the two sets were the same size. And he was able to do so using the following surprisingly simple argument.

Consider any point (x, y) in the unit square, say (.364, .283). We now associate it with a point on the unit line segment by interspersing the digits of the two numbers x and y. Our new number uses the first digit of x, then the first digit of y, then the second digit of x, then second digit of y, and so on. Our new number is .326843, constructed from all the digits of both x and y. Note that this construction works even if x and y are of infinite length. Cantor pointed out that in this

[8]Emmy Noether and Jean Cavaillès, eds., *Cantor-Dedekind Briefwechsel* (Paris: Hermann, 1937), 20–21. English trans. in William Bragg Ewald, *From Kant to Hilbert*, vol. 2 (Oxford: Oxford University Press, 2007).

way any (x, y) pair—a point in the two-dimensional unit square—was assigned a unique point on the unit line segment.

Similarly, any number on the unit line segment is assigned back to its own (x, y) pair in the square by simply reversing the process. Let the first digit of the number become the first digit of x, the second digit of the number become the first digit of y, and so on. In this way, .326843 is assigned to (.364, .283). Using this construction, Cantor had discovered a way to find a one-to-one correspondence between the sets, leading him to conclude that the set of points on the unit line segment and the set of points in the unit square are of the same cardinality.

Cantor was dumbfounded, writing in a letter to a friend and colleague, "I see it, but I don't believe it!"[9,10] And how could he? Intuitively, it's difficult to reconcile that a square and a line segment could possibly contain the same number of points. Yet in a well-defined sense, they do. Cantor had (literally) crossed a line mathematicians never imagined possible.

Cantor's result was emblematic of a broader theme in mathematics that was transpiring during the late nineteenth and early twentieth centuries: a better understanding of infinity as it related to the foundations of math. Mathematicians realized that unless they were careful when dealing with infinity, they could arrive at absurd results—Grandi's "proof" that $0 = 1/2 = 1$ being a simple yet illustrative example. And infinity couldn't be brushed aside since most of mathematics dealt with infinity in one form or another. So various efforts were undertaken to provide a simple yet rigorous foundation for dealing with infinity—the concept of limits being a major part of

[9] In fact, Cantor was a bit premature in his proclamation. The proof described here contains a subtle error pointed out by Richard Dedekind. A discussion of the error and a means by which to correct it are described in the appendix to this chapter.

[10] David Foster Wallace, *Everything and More: A Compact History of Infinity* (New York: W. W. Norton, 2003), 259.

that effort. Yet, even as mathematicians attempted to shore things up, perplexing peculiarities continued to present themselves.

One of the most historically significant was pointed out by Bertrand Russell in 1901. Until that time, mathematicians had typically treated the notion of a set informally. A set was a collection of "things." The teacups on a table. The real numbers. The set of all sets that contain a teacup.

Russell asked us to consider the following set: the set S of all sets that don't contain themselves as members. We now ask the following question: is S contained in itself? We first observe that S can't contain itself by virtue of the way we defined S (if it contained itself, it wouldn't be a set that didn't contain itself, and therefore wouldn't be in S by definition). Therefore, S doesn't contain itself. But if it doesn't contain itself, it must contain itself, again by the way we defined S. The set S is self-contradictory.

The set S wasn't so much a paradox as it was a call to action. Mathematicians needed to clean up the rules of how we defined and worked with sets so that contradictions like Russell's couldn't occur. The task was complicated because it represented a balancing act. On one hand, the rules needed to be restrictive enough to eliminate contradictions. On the other, they needed to be rich enough to encompass the realm of mathematics that mathematicians wanted to explore. For example, the rules could state that "a set can only contain a finite number of elements," but this wouldn't allow investigation of the real numbers or even the counting numbers. In the other direction, in developing his many levels of infinity, Cantor had made use of "the set of all subsets of an infinite set." Was he free to do so, or did this lead to some hidden contradiction like Russell's set?

The search for the best set of rules, or axioms, on which to build set theory and, by extension, modern mathematics continues to this day. Important advances made over the years have led mathematicians to largely focus on a handful of related collections of axioms. Yet it seems that any collection of axioms rich enough to meet the needs of the mathematical community comes with seemingly paradoxical peculiarities.

Cantor's results, which remain true under today's most frequently adopted axiomatizations, are among these peculiarities. It seems difficult to imagine that there's more than one kind of infinity. The thought that there are an infinite number of infinities, each infinitely larger than the next, is startling, to say the least. Yet it poses no logical contradictions when starting from first principles.

One of the earliest mathematical peculiarities arose from the work of Karl Weierstrass, one of Cantor's teachers. Weierstrass devoted much of his career to rigorously establishing the foundations of math. Among other things, he sought to formalize and clearly establish some of the most basic concepts in all of mathematics—concepts mathematicians worked with daily.

One of these concepts was *continuity*, which students learn early on. Figure 4-5 and Figure 4-6 show examples of continuous and discontinuous functions, respectively. A simple and frequently adequate definition of a continuous function is a function you can draw without picking up your pencil. If Figure 4-5 represents the speed of a car accelerating to 60 miles per hour, we'd expect it to pass through all velocities between 0 and 60 along the way. There's no point at which it magically jumps from 30 to 40 miles per hour instantaneously as in Figure 4-6. The function might be steep if the car is accelerating quickly (a Tesla) or gradual if the car is accelerating slowly (a Prius). But in neither case does the car skip any of the intervening velocities. We note that the function in Figure 4-6 is almost continuous in that there is only one point of discontinuity. Continuous functions are considered important, and in many ways the most natural functions, because they represent how we expect most real-world phenomena to behave.

Another concept addressed by Weierstrass was that of *differentiability*. Differentiability has to do with the lack of "kinks" in the function. Figure 4-7 and Figure 4-8 show examples of differentiable and

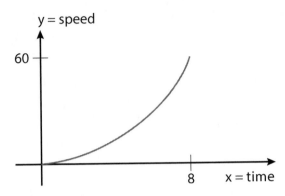

Figure 4-5: A continuous function

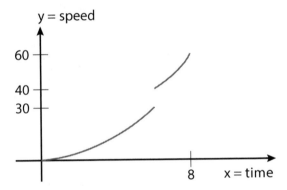

Figure 4-6: A discontinuous function

nondifferentiable functions, respectively. Loosely speaking, a curve is differentiable at a point *x* if the slope is smooth at that point. We note that the function in Figure 4-8 is almost differentiable in that there is only one kink in the slope. Differentiable functions are considered as basic as continuous functions since they make the same statement about changing acceleration as continuity makes about changing velocity. Weierstrass provided formal definitions of both continuity and differentiability using the notion of limits.

It's not difficult to imagine drawing a continuous function with many kinks, like the function in Figure 4-9, though at a certain point the effort would become tedious. But Weierstrass asked himself a more difficult question. Can I find a function that's con-

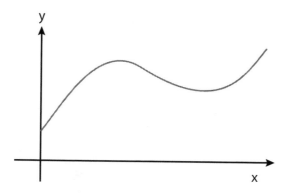

Figure 4-7: A differentiable function

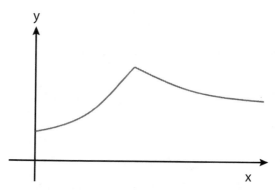

Figure 4-8: A non-differentiable function

tinuous everywhere but differentiable nowhere? A function that can be drawn without lifting pencil from paper that has a kink at every point?

To help emphasize how astonishing this state of affairs would be, it's worth taking a closer look at the real numbers by asking a basic question: what's the next biggest real number after 0?

Suppose I claimed I had such a number. By taking that number and dividing it in half, we'd have an even smaller number that was still bigger than 0. From this we realize there is no next biggest real number after 0.

Now imagine drawing a continuous function—any continuous function—starting at $x = 0$ and moving to the right—that has a kink at every

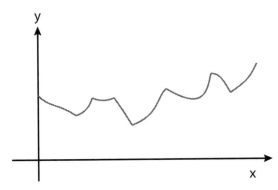

Figure 4-9: A continuous function with 7 non-differentiable points

real number.[11] How do we begin to even imagine what such a function looks like, much less draw it, since the "next" number to the right doesn't exist? Armed with infinity, however, we know that such functions exist as demonstrated by Weierstrass. One such function is

$$f(x) = \sum_{n=0}^{\infty} \left(\frac{1}{2}\right)^n \cos\left(13^n \pi x\right)$$

The function corresponding to the first term of the summation, $\cos(\pi x)$, is plotted in Figure 4-10. Observe that the function shown in the figure is itself both continuous and differentiable, as are all the individual terms in the summation. Yet somehow, when we add them together, the aggregate function is nowhere differentiable. We can't draw the function since it's defined as the summation of an infinite number of terms. However, working from the definitions proposed by Weierstrass, definitions taught to students the world over when introduced to calculus, we can show there exist a vast number of continuous but nondifferentiable functions. In fact, it's been shown in a well-defined mathematical sense that these peculiar functions vastly outnumber the continuous, differentiable variety mathematicians find in the real world.

---

[11] It's yet another version of Zeno's paradox to realize that we can, in fact, draw such a function even though there's no next number greater than 0.

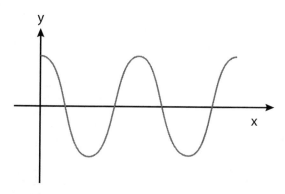

Figure 4-10: The first term in the Weierstrass summation

In setting about to tame infinity by laying out simple, concise definitions and axioms, mathematicians soon realized that infinity held more surprises than they could have imagined. In 1899, Henri Poincaré summarized the situation.

Logic sometimes makes monsters. For half a century we have seen a mass of bizarre functions which appear to be forced to resemble as little as possible honest functions which serve some purpose. More of continuity, or less of continuity, more derivatives, and so forth. Indeed, from the point of view of logic, these strange functions are the most general; on the other hand those which one meets without searching for them, and which follow simple laws appear as a particular case which does not amount to more than a small corner.[12]

While not all mathematicians agreed with Poincaré's characterization that "logic sometimes makes monsters," they did agree the mathematical world was looking far stranger than expected. And as they explored the boundaries of infinity ever more closely, their beliefs were only reaffirmed. Math is filled with results profoundly counter to our intuition. We accept them because they logically arise from more basic,

---

[12] William Bragg Ewald, *From Kant to Hilbert, Volume 2* (Oxford: Oxford University Press, 2007), 973.

more intuitive principles that, when pushed to their limit,[13] give rise to bizarre offspring—offspring that challenge our very ability to comprehend them.

References and Further Reading:

1. Dauben, Joseph Warren. *Georg Cantor and the Battle for Transfinite Set Theory*. New York: CUNY, 1988.
2. Dauben, Joseph Warren. *Georg Cantor: His Mathematics and Philosophy of the Infinite*. Princeton, NJ: Princeton University Press, 1990.
3. Edwald, William Bragg. *From Kant to Hilbert, Volume 2*. Oxford: Oxford University Press, 2007.
4. Kline, Morris. *Mathematical Thought from Ancient to Modern Times*. Vol 3. New York: Oxford University Press, 1972. See in particular Section 7, "The Status of Analysis," in Chapter 40, "The Instillation of Rigor in Analysis," 972-977.
5. O'Connor, J. J. and E. F. Robertson, "Georg Ferdinand Ludwig Philipp Cantor." From the MacTutor History of Mathematics Archive, School of Mathematics and Statistics, University of Saint Andrews, Scotland. Accessed October 19, 2015.
6. Wallace, David Foster. *Everything and More: A Compact History of Infinity*. New York: W. W. Norton, 2003.

---

[13]A more accurate phrase might be "pushed to infinity."

Appendix

A "Proof" That $1 + 2 + 3 + 4 + \ldots = -1/12$

Begin by considering the slightly different summation

$$S = 1 - 2 + 3 - 4 + \ldots$$

Add this number to itself, but do so after shifting the second copy to the right by one.

$$
\begin{array}{rrrrrr}
 & 1 - 2 + 3 - 4 + 5 -\ldots \\
+ & \quad\; 1 - 2 + 3 - 4 +\ldots \\
\hline
 & 1 - 1 + 1 - 1 + 1 -\ldots
\end{array}
$$

So $S + S = 2S = 1 - 1 + 1 - 1 + 1\ldots$

Recall that, using long division, Grandi argued that the series $1 - 1 + 1 - 1 + 1\ldots$ had a value of $1/2$ (and other values; we adopt $1/2$ for the remainder of the discussion). Thus, $2S = 1/2$, or equivalently, $S = 1/4$ (which is already clearly incorrect).

To continue, call our original summation of interest T, so $T = 1 + 2 + 3 + 4 +\ldots$, and calculate $T - S$.

$$
\begin{array}{rrrrr}
 & 1 + 2 + 3 + 4 + 5 + 6 +\ldots \\
- & (1 - 2 + 3 - 4 + 5 - 6 +\ldots) \\
\hline
 & 4 + \quad 8 + \quad 12 +\ldots
\end{array}
$$

Notice that this difference can be written as

$$4(1 + 2 + 3 + 4 +\ldots) = 4T$$

Summarizing, $T - S = 4T$, or rearranging terms, $T = -(1/3)S$. But above we argued $S = 1/4$. Therefore,

$$T = 1 + 2 + 3 + 4 +\ldots = -(1/3)S = -(1/3)(1/4) = -1/12$$

The Problem With Cantor's Proof that the Set of Points in the Unit Line Segment and in the Unit Square are in One-to-One Correspondence

In order to understand the nature of Cantor's error, recall that we argued the number 1 can be represented in two different ways: as the repeating decimal .99999... or as the single digit 1. Using similar logic, it can be argued that any finite decimal fraction can be written in two ways, one of which involves an infinite sequence of repeating nines. For example, 1.75 = 1.7499999... and 13.2 = 13.1999999...

Consider the following two numbers taken from the unit line segment.

$$0.2101010101...$$
$$0.1191919191...$$

When breaking out (x, y) pairs in the square using Cantor's construction, we find

$$(x, y) = (.200000..., .111111...)$$
$$(x, y) = (.199999..., .111111...)$$

Observe that these last two (x, y) pairs represent the same number. Based on Cantor's construction, each number in the unit square corresponds to more than one number on the unit line segment. Therefore, the correspondence isn't one-to-one, though it does show that the unit line segment has at least as many points as the unit square. Cantor might have contented himself with this result, but he strongly desired to show the two sets were of the same cardinality, and to do this he needed a true one-to-one correspondence. He managed his way through such a proof, but a truly elegant resolution would wait ten years for a result known as the Schröder-Bernstein theorem.

It's a good idea to ask ourselves, who do we think we are?

—John Searle

Chapter Five
Consciousness

As you read this sentence, what are you thinking? It probably has something to do with the sentence. But what's more important is that there's no denying you're thinking something.

Consciousness is that sequence of thoughts you and I experience during our waking lives. It's so natural, so persistent, we take it for granted. One thought follows the next from moment to moment. That doesn't occur when we sleep. At some point, as we lie in bed, we relinquish conscious thought. Our brains and bodies remain quite active when we sleep, but our sense of awareness is absent until we awaken.

Our thought process is frequently intentional. I might think, "I want to stand up, go to the kitchen, and get a glass of water," and then do so. Sometimes it isn't intentional, as when I stare out the window and allow my mind to drift from one idea to the next—from the sound of the water flowing by, to thoughts of fishing, to thoughts about what will be for dinner this evening.

Consciousness also carries with it inner feelings and sensations. "I'll get a glass of water" is a thought. But it's not the experience I feel when I drink the water following a long run in hot weather. Likewise, the pain I feel when I set my hand on a hot stove and the awe I feel when watching the stars on a clear evening are parts of my conscious experience.

Philosophers have long asked themselves about consciousness—what it is and what it isn't. More recently, with the advance of technology, ever more fields of study are entering the debate. Medical researchers probe and map the brain in an effort to determine what regions are related to particular thoughts and actions. Physicists raise questions about how quantum theory might be involved. Computer scientists ask if it will be possible to replicate human thought and, if so, if computers will develop consciousness.

The fact that so many people are studying the capabilities of the human brain is no surprise. Doctors hope to develop new ways to help people with brain disorders, and viewed solely as a computational device, the brain is capable of performing tasks even our most sophisticated computers and algorithms have yet to accomplish. We will undoubtedly reap the benefits of these efforts. But for all our effort and creativity, there remains a fundamental, paradoxical question: how does the mind arise from the brain? How do lifeless bits of matter give rise to consciousness, complete with all the inner sensations that go along with it? It's beyond comprehension.

To drive home the point, imagine the following. You're provided with the following list of ingredients:

> 97 1/2 lb. oxygen
> 27 lb. carbon
> 15 lb. hydrogen
> 4 1/2 lb. nitrogen
> 2 1/4 lb. calcium
> 1 1/2 lb. phosphorous
> 1/2 lb. potassium
> 3/8 lb. sulfur

With these ingredients, you now have over 99 percent of what it takes to make a 150-pound human. (The small amounts of additional ingredients aren't listed, but they're available as well.) The fact is inescapable. We're made up of the same stuff as rocks, celery, and toaster ovens. It's objectively unreasonable that these lifeless materials can be combined in a way that gives rise not only to life but to our remarkable human life with human consciousness. Yet we know from irrefutable firsthand knowledge that such materials can give rise to conscious life— we're living proof.

Surely modern research into the brain will answer this difficult question. Or will it? Consider, for example, the intense research focused on mapping the brain. One popular form of seeing what goes on in the brain is functional magnetic resonance imaging, or fMRI. It's known that the activity of the brain's neurons is related to blood flow, which carries oxygen. An fMRI can track changes in blood flow and therefore changes in neuronal activity. After placing a subject's head in a magnetic resonance imaging device, a simple experiment might consist of showing the subject pictures of people in different social situations. An fMRI creates three-dimensional pictures of the brain using colors to show changes in the level of blood flow. In this way, researchers are able to determine what parts of the brain show activity when, for example, a subject sees a picture of a face.

The use of fMRI is just one way in which researchers are seeking to understand the mechanics of brain activity, with the emphasis on mechanics. What fMRI can't do is explain why certain mechanical responses to standing beneath a waterfall translate to a conscious sensation of calm (or whatever inner sensation is evoked)—or even more fundamentally, what gives rise to consciousness at all.

Modern science is capable of tackling mechanical questions similar to the questions being asked using fMRI. Hypothesizing that inadequate neuronal activity in a particular region of the brain is a sign of a neurological disorder is a perfectly valid line of scientific inquiry. An attempt to scientifically explain consciousness, however, is a different animal.

*****

Our dog, Horace, barks at the postal carrier. Every day as the mail slides through the slot in the door, Horace runs to the door, barks, then runs to a nearby window, where he continues to bark as the carrier makes his way down the street. We've scolded and cajoled. We've even introduced Horace to the carrier in the hope that familiarity will fix the problem. Still, the barking and running around continue.

It's been explained to me that every visit by the postal carrier is a new experience for Horace—that he doesn't retain memories of the past. He wants to protect us, and he protects us from what he perceives as a new and different danger with every clank of the mail-slot door. It seems plausible, but it's not entirely convincing. Horace recognizes family and friends following long periods of absence. Like most dogs, he remembers and responds to certain words. He also knows where the food is kept. Horace has demonstrated that he has the capability to learn and to use that knowledge to navigate the world about him. So why does he insist on barking at the mail carrier day after day when every effort's been made to teach him not to? What's going on inside his head?

The fact is I don't know what's going on inside his head, and I never can. I can see how Horace reacts to various situations, but I can never *be* Horace. I can't magically place myself in his head and experience what it's like when he senses the mail carrier at the door, or when he sees a familiar face, or when he hears the word "no." Horace is Horace, with his own set of personal experiences and sensations that I have no access to.

Armed with this observation, imagine I claim to have discovered the cause of consciousness. My hypothesis could be as simple as the existence of a particular neurotransmitter or as complicated as a diagram outlining millions of neural connections and how they interact over time. The specifics of my hypothesis aren't important, only that I can point to a particular, scientifically verifiable set of conditions. My claim is that when the conditions are present, an entity is conscious, and when they're not, an entity isn't. How do I prove my hypothesis is correct?

It's important to keep in mind that the hypothesis isn't about *behaving* consciously but about actually *being conscious*. We can imagine a robot that behaves like a human (or dog or other conscious entity) without the robot actually being conscious. Robots that behave like humans are a recurrent theme in science fiction. Philosophers refer to hypothetical entities that act like humans but don't experience human sensation as (philosophical) zombies—a technical term. Acting conscious—behaving the way a conscious entity would behave—and actually being conscious are two different things.

So looking at my hypothesized conditions for consciousness, how do we determine if they do indeed lead to consciousness? To do so, we'd need to "get inside the head" of the entity to see if yes, when these conditions are present, the entity actually experiences consciousness, and no, when they're not present, the entity isn't conscious. But in our discussions of Horace we concluded that this is impossible—that there's no way to experience the inner sensations felt by another entity. As a result, a set of scientific conditions that form the basis for consciousness can never be discovered and validated scientifically.

The issue isn't one of technology but of the question being asked: what conditions cause consciousness? Asking if an entity behaves in a conscious fashion is easy. Suppose we're confronted with a mechanical dog whose owner claims that it's conscious thanks to a new microprocessor based on a revolutionary new technology unlike anything used before. We can certainly observe the mechanical dog to see if it behaves like a real dog. We'd likely argue over the details. If the mechanical dog fails to chase a ball when it's thrown, is the response consistent with that of a real dog? The point, however, isn't the details; it's that we can have a perfectly valid discussion about what behaviors are consistent with being a dog. And we can observe the mechanical dog to see what behaviors it exhibits and if these behaviors match what we believe to be those of a real, bona fide, in-the-flesh dog.

The question we're interested in, however, is not if the mechanical dog behaves as if it's conscious, but whether it is, in fact, conscious. When the mechanical dog climbs on the couch and places its head on its owner's lap, does it feel content or happy or safe or anything else? Or is the activity devoid of inner sensation, akin to what my room-roaming automatic vacuum feels as it picks up dog hair? There's no way of knowing, since we have no way to experience those inner sensations. Hence we can't scientifically verify that the new microprocessor has given rise to consciousness.

Which leaves us further mired in a conundrum. Not only is it objectively unreasonable for life to arise from lifeless materials—even though it clearly does—but we can't hope to make a meaningful scientific statement about how this actually occurs. Conscious life arising from lifeless bits of stuff is beyond comprehension.

*****

In 1982, philosopher Frank Jackson introduced us to Mary, a capable, hardworking scientist who for some unknown reason is locked in a room without color. Everything appears in various shades of gray. Mary isn't real, of course; she's the basis of a thought experiment Jackson uses to make a point about consciousness.

Mary has chosen to study vision. From within her room, on her black-and-white computer, Mary studies the neurophysiology of vision. She understands, for example, that the color "red" is a certain wavelength of light, that when light of a particular wavelength strikes someone's retina it sends signals to the brain, and so on. She becomes an expert, learning every physical, scientific fact about light and vision.

Upon completion of her studies, Mary is allowed to leave her black-and-white room, and for the first time she experiences red and blue and all the other colors of the world. In a word, her reaction is likely to be "wow." Jackson's example powerfully illustrates the experiential nature of consciousness—the fact that consciousness is more than mere thinking. It includes the many sensations experienced by the conscious mind.

But Jackson is making a bigger point, a point that casts further light upon the inexplicable nature of consciousness. He argues that when Mary leaves the room, she learns something new—what it's like to actually experience red, blue, and all the other colors. As such, she learned something she wasn't able to ascertain just by studying the physical world. Ergo, not everything that can be known can be learned by studying the physical world.

It's a tight, simple argument that's worth dwelling upon. Mary has access to all the physical facts that can be known about vision. She understands the world of light waves and molecules and chemical reactions and so on. Yet when she sees color for the first time, she learns something new: "Oh, *that's* what it's like to see red." What she learns is so new, so different from anything she knew before, that she finds it hard to contain herself. She experiences one of those moments of learning that teachers spend their lives seeking to impart to their students.

But the learning didn't come from anything new she learned about the physical world. The learning came from an inner, qualitative experience. Her eyes gazed upon red, and she knew something she hadn't known from all her studies of the physical world. Her knowledge prior to this experience was incomplete. Therefore she's learned something that couldn't have been knowable just by studying the physical world.

The conclusion is startling and invites the question: if there exists knowledge that doesn't have its basis in the physical world, what's the basis of this knowledge? From where does it come?

The answer that leaps out is that the knowledge comes from some-where apart from the physical world. Consciousness, along with many or all of its properties, relies on the existence of an invisible realm. This position has the advantage that the quandary of inanimate materials giving rise to consciousness goes away. How do our lifeless ingredients think and experience inner sensations? Because, it would be argued, consciousness doesn't depend on the material substance of our brains. Our brains may be part of what enables consciousness, but there's something lurking in the background that animates us—something that gives more to life than oxygen, carbon, hydrogen, and all the other ele-ments that make up the human body.

The obvious disadvantage of this position is that the question it raises is as bad, or worse, than the answer it provides. What, exactly, is this non-material thing that animates us, that provides the basis of consciousness? It's mysterious. How do we know what's true and what's false about this nonphysical domain? It's possible to claim virtually anything about it.

But most philosophers would, in addition, offer an even simpler objection: get it through your head—the physical world is it. Don't go dragging mysterious, unobservable domains into the argument. Propo-nents of this "what you see is what you get" position argue that the body is exactly the collection of material ingredients listed earlier. If we are to explain consciousness, we must explain it in these physical terms. But then how do we explain Mary's new knowledge, knowledge that didn't result from study of the physical world?

The question is how to reconcile these two positions. On one hand, the argument from Mary's room leads to the conclusion that not every-thing can be learned by studying the physical world; there exist dual realms, one physical, one not. On the other hand, if we accept the exis-tence of knowledge that transcends study of the real, physical, material world, we've conceded the existence of something mysterious.

The gap is naturally bridged if we accept the Beyond Comprehen-sion (BC) conjecture: there are real things in our world that humans can't comprehend. Mary's knowledge does, in fact, come from the real, physical world, but it derives from something we aren't wired to grasp. The physical world isn't abandoned in favor of the mysterious. We simply concede there are parts of the physical world we can't access.

It's important to emphasize that reconciling these two positions in this way isn't simply a matter of semantics. Human history is filled with references to nonphysical worlds, from Plato's realm of ideals to the spiritual realms espoused by Christianity, Islam, and other religions. Purveyors of these realms make no claim they're physical in the same sense as the physical world in which we live. In fact, references to these nonphysical realms propound a distinctly different reality. When Christians point to heaven, it's not something found in our universe. And the mystery affiliated with heaven is not only acceptable, it's presumed. On the other hand, those who reject this standpoint and defend the position that there exists only a real, physical, material world mean exactly that: a real, physical, material world.

The disagreement between the positions is fundamentally about whether some altogether different reality exists apart from our physical world. The BC conjecture postulates no such division. It places everything squarely in the physical world while simultaneously recognizing that, yes, there may be more to life than the lifeless ingredients our senses are familiar with. As such, the conjecture allows us to accept that there may be more to our world than meets the eye while at the same time placing clear limits on what we can rationally claim. Mary learned something when she left her colorless room. She discovered the experience of actually seeing red. How can that be? We don't know, and we don't claim to know. It's beyond comprehension.

*****

Before leaving the topic of consciousness, it's worth noting that ideas found in quantum theory have been proposed as offering clues to understanding human consciousness. Physician Stuart Hameroff and physicist

Roger Penrose go so far as to suggest specific structures within the brain that may actually serve as conduits for quantum activity leading to consciousness. The results are highly speculative and open to broad criticism. Hameroff and Penrose deserve credit, however, for seeking to make a novel connection between consciousness and a respectable field of scientific study. But does thrusting the question of consciousness into the quantum arena buy anything? At best, purveyors of quantum consciousness theories may simply be replacing one incomprehensibility with another. Philosopher Patricia Smith Churchland has gone so far as to comment that "Pixie dust in the synapses is about as explanatorily powerful as quantum coherence in the microtubules."[1]

The difference between pixie dust and quantum coherence, however, is substantial. While quantum theory is a wellspring of befuddlement, we can deal with it scientifically. Quantum computing, though still in its infancy, is real. If Hameroff, Penrose, and like-minded individuals prove to be correct (and it's again worth emphasizing that this isn't at all clear), the results would surely be interesting. They might even lead to medical advances. But even these results couldn't overcome the basic problem that no scientific conditions claiming to be the basis for consciousness, even those based on something as unusual as quantum theory, can be discovered and validated scientifically.

References and Further Reading

1. Churchland, Patricia Smith. *Brain-Wise: Studies in Neurophilosophy*. Cambridge, MA: MIT Press, 2002.
2. Hameroff, Stuart. "Quantum Coherence in Microtubules: A Neural Basis for Emergent Consciousness." *Journal of Consciousness Studies* 1:1 (1994): 91–118.

---

[1]Patricia Smith Churchland, *Brain-Wise: Studies in Neurophilosophy* (Cambridge, MA: MIT Press, 2002), 197.

3. Hameroff, Stuart and Roger Penrose. "Conscious Events as Orchestrated Spacetime Selections." *Journal of Consciousness Studies* 3:1 (1996): 36–53.

4. ———. "Orchestrated Reduction of Quantum Coherence in Brain Microtubules: A Model for Consciousness." In *Toward a Science of Consciousness: The First Tucson Discussions and Debates*, edited by Stuart Hameroff, Alfred Kaszniak, and A. C. Scott. Cambridge, MA: MIT Press, 1996.

5. ———. "Consciousness in the Universe: A Review of the 'Orch OR' Theory." *Physics of Life Reviews* 11:1 (2014): 39–78.

6. Jackson, Frank. "Epiphenomenal Qualia." *The Philosophical Quarterly* 32:127 (1982): 127–36.

7. Ludlow, Peter, Yujin Nagasawa, and Daniel Stoljar, eds. *There's Something About Mary: Essays on Phenomenal Consciousness and Frank Jackson's Knowledge Argument.* Cambridge, MA: MIT Press, 2004.

8. Nagel, Thomas. *Mind and Cosmos: Why the Materialist Neo-Darwinian Conception of Nature is Almost Certainly False.* Oxford: Oxford University Press, 2012.

9. ———. "What Is It Like to Be a Bat?" *The Philosophical Review* 83:4 (1974): 435–50.

10. Penrose, Roger. *The Emperor's New Mind: Concerning Computers, Minds, and the Laws of Physics.* New York: Penguin, 1989.

11. Searle, John. *Mind: A Brief Introduction.* Oxford: Oxford University Press, 2004.

12. ———. *The Mystery of Consciousness.* New York: New York Review of Books, 1997.

13. ———. *The Rediscovery of the Mind.* Cambridge, MA: MIT Press, 1992.

I think I can safely say that nobody understands quantum mechanics.

—Richard Feynman

CHAPTER SIX
Quantum Weirdness

Newton was unquestionably one of history's most creative minds, but he wasn't always right. With his reputation firmly established by his work on gravity, Newton's work in other areas was held in high esteem. Anyone seeking to contradict the great master faced an uphill battle. Still, should someone come forward with strong enough evidence to contradict one of Newton's theories, scientific minds would of necessity have to reconsider that theory.

On November 24, 1804, a century after the publication of Newton's great treatise *Opticks*, Thomas Young stood before the Royal Society of London professing to have such evidence. Young was born in 1773 in Milverton, England, the eldest of ten children. By age fourteen, he'd learned Greek and Latin and was familiar with a dozen more languages. At just twenty-one, he was appointed a member of England's Royal Society; at twenty-three, he earned a degree in medicine and began work as a physician. A few years later, Young was appointed a lecturer in natural philosophy at the prestigious Royal Institution of England. When pressed by the *Encyclopaedia Britannica* for contributions, Young offered his expertise on a long list of topics ranging from annuities to tides. So broad were his interests that one biographer was moved to call Young "the last man who knew everything."[1]

In *Opticks* Newton put forward the theory that light was a particle. Young had evidence it was a wave. As Young expressed in a presentation to the Royal Society, "the assertion is proved by the experiments I am about to relate, which may be repeated with great ease, whenever the sun shines, and without any other apparatus than is at hand to everyone."[2] To better appreciate the experiment, it's first worth taking a closer look at the properties of particles and waves.

A bullet is an example of a particle. While we typically reserve the word "particle" for very small entities, the essence of a particle is that it has mass and takes up space. When a moving particle hits an object, it imparts energy. If I shoot a bullet at a wall, the wall takes the brunt of the energy contained in the bullet at the moment of impact.

Suppose we used bullets to perform the following experiment, depicted in Figure 6-1. In the middle of a room is a barrier with a slit cut into it wide enough to accommodate the passage of a bullet. Someone with a rifle stands on one side of the room and shoots randomly at the barrier. Most of the bullets bounce off, but a few make it through the slit and hit the wall on the far side of the room. What shape indentation do these bullets form?

If we shoot enough bullets, the indentation should begin to take the shape of the slit. And whatever the details of what we see, every individual bullet hole should be in the line of sight of the rifle. Bullets don't magically turn corners. They travel in straight lines.

If light was a particle, we'd expect it to behave in much the same way as bullets. However, as Young discovered, under the right condi-

[1]Andrew Robinson, *The Last Man Who New Everything: Thomas Young, the Anonymous Polymath Who Proved Newton Wrong, Explained How We See, Cured the Sick and Deciphered the Rosetta Stone* (London: Oneworld Publications, 2007).

[2]Thomas Young, "The Bakerian Lecture: On the Theory of Light and Colors," *Philosophical Transactions of the Royal Society of London* 92 (1802): 12-48.

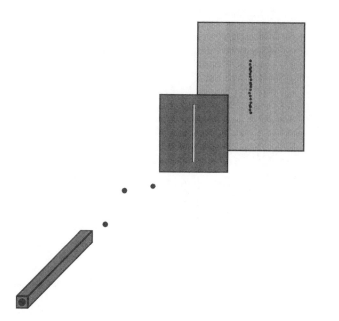

Figure 6-1: Bullets shot through a slit in a barrier

tions we can observe that light doesn't behave like bullets at all. But it does behave like waves.

Waves are a bit more complicated than particles. Waves carry energy, but without moving any particles from point A to point B. When we drop a stone in a pond, waves emanate that carry energy capable of moving sand on the beach. But no water molecules have actually moved from the point of the stone's entry to the shore. Instead, the water molecules engage in a choreographed pushing of one another. This pushing allows energy to be transferred through the water until it reaches the shore. Observe, as in Figure 6-2, that the wave crests rise above the waterline while the wave troughs lie below the waterline. And the waves smoothly wash up onto the beach over a period of time. There's not a single thud, as when a bullet strikes a wall.

Waves come in different types (sound waves, for example, result from a different type of pushing than waves on the surface of water), but they all share certain characteristics. One of these characteristics

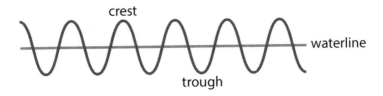

crest

waterline

trough

Figure 6-2: A simple wave

is *diffraction*—the way waves behave when they encounter a barrier or other obstacle in their path.

To illustrate diffraction, consider the water waves shown in Figure 6-3. Here, a long board is pushed back and forth in the water, creating identical, parallel waves moving from left to right until they encounter a barrier with a slit. Due to the way in which the waves are generated, they are *coherent* in that the distance from one crest to another is the same over the entire wave. We can observe properties in coherent waves that we can't in less orderly waves.

Of interest is that the wave radiates outward in a circular fashion as it passes through the slit. Mathematical theories have been developed that explain this diffraction pattern, but for our purposes, we simply take it as given that this is, in fact, the way waves propagate when they encounter a slit.[3]

One of the interesting artifacts about the way waves interact with slits is that waves aren't limited to line-of-sight travel in the same way as bullets. Waves can and do turn corners, as we know from the fact that we can hear someone in another room even if we can't see them. But the height of the wave, as it moves forward from the slit, is no longer uniform but varies depending on the angle with the slit.

Consider now the intensity-detection device in Figure 6-3. Since the waves are in motion, we would expect their height to change over time. So rather than measure the height at one instant, we'll measure it

---

[3]This fact is easily verified in a water tank, and many good websites exist that demonstrate this phenomenon and explain it in greater detail.

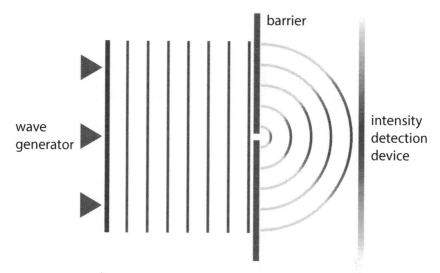

Figure 6-3: Water waves diffracting as they
pass through a slit in a barrier

over time and take an average. In fact, we won't average the height, but
the square of the distance between the wave's height and the waterline.
Why? Because this last measure is proportional to the wave's intensity—
both in a formal mathematical sense and in everyday usage.[4]

As might be expected, the detector shows that waves are most
intense directly behind the slit, and the intensity falls off symmetrically
to both the right and left. The pattern is actually somewhat more complicated than this, but for a sufficiently thin slit, the pattern depicted
here is a good approximation.

We now have the makings for an experiment to determine if light is
a particle or a wave. Shine a coherent light source through a thin slit. If
light is a particle, like the bullets, we should see something that resembles the shape of the slit on the back wall. If light is a wave, we should
see a light intensity pattern that fades to the right and left of the slit.
The two possibilities are depicted in Figure 6-4.

[4] It also yields a nonnegative intensity for negative wave heights, and the intensity is the same for identical deviations, be they above or below the waterline.
The square of the distance is used rather than the absolute value for technical
reasons; among them, it fits with the established formal definition of intensity.

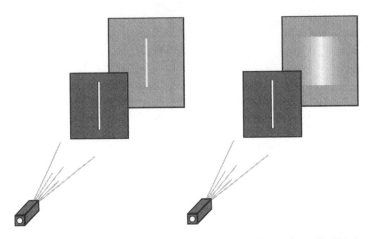

Figure 6-4: Light as it would appear through a slit if it is
a particle (left) or a wave (right)

With today's laboratory instruments, it's easy to carry out this experiment. In Young's day, however, the experimental means available would have made it considerably more difficult to determine if the light looked like a slit or a fuzzy slit. So Young relied on a variation of the experiment that was easier to verify with the naked eye. His experiment relied on another characteristic common to waves, *interference—* how waves act when they encounter other waves.

To illustrate wave interference, we introduce a second slit, as depicted in Figure 6-5. Note that the waves emanating from both slits individually are identical to those depicted in Figure 6-3. Of interest is the fact that when two waves meet, the height of the combined wave is equal to the sum of the heights of the individual waves. An example is shown in Figure 6-6.

We're now prepared to take a closer look at the waves generated, as shown in Figure 6-5. Using the same measuring device as before, we can determine the average intensity of the combined wave at each point where it strikes the back wall. The figure illustrates a key advantage of using two slits: waves reach peak intensities at many different points along the measuring device. These peaks are interspersed with points of zero intensity where the waves arriving from the two different slits cancel each other out. The intensity detection device displays the *interference pattern* of the two waves.

We now have a more easily interpretable experiment for examining whether light is a wave or a particle. Shine a coherent light source through two thin, parallel slits. If light is a particle, we should expect to see something that resembles the shape of two parallel slits on the back wall. If light is a wave, we should see a distinctive interference pattern. The two possibilities are depicted in Figure 6-7.

When Young walked into the chamber to speak to the Royal Society, he first demonstrated interference and diffraction in water waves using a wave tank he'd put together. He then presented the double-slit experiment using not water but light. On a thin glass plate covered in soot, he cleaned two thin lines parallel to one another. When an appropriate source of light was cast upon the plate, the unmistakable bands of wave interference were visible. It would seem that light couldn't be a particle. It must be a wave.

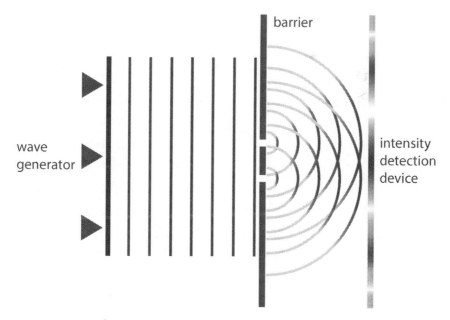

Figure 6-5: Water waves interfering as they pass through two slits in a barrier

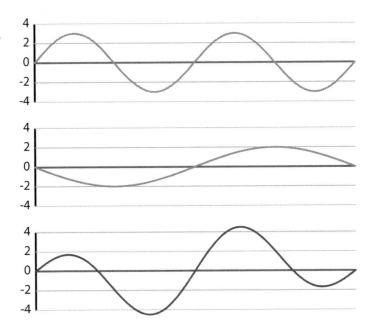

Figure 6-6: When two waves meet, the height of the
combined wave is their sum

With such strong evidence, it might be expected that the Royal
Society would be dumbfounded. But that wasn't the case. The long-
held belief that light was a particle wasn't going to go quietly. And
even though the double-slit experiment would have been easier to
verify than the single-slit version, it still had limitations. To obtain
a nearly coherent light source, Young used sunlight filtered through
a single slit. This was sufficient to see the two-slit diffraction pattern
but at the cost of diminishing the intensity of the light that fell on the
two slits and in turn making the pattern harder to see. Today, with
the aid of a laser pointer and tape, the experiment can be performed
with vivid clarity.

To understand what happened, it's worth recognizing that while
many properties distinguish waves from particles, two in particular
stand out.

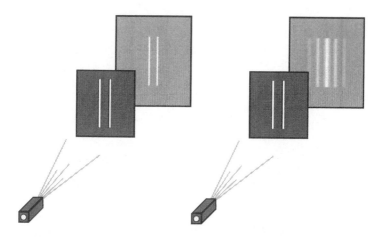

Figure 6-7: Light as it would appear through two slits if it is a particle (left) or a wave (right)

1. Waves turn corners when they encounter barriers. Particles don't.

2. Waves transfer energy little by little over time. The longer ocean waves lap upon the shore, the more energy they impart. Particles, like billiard balls, transfer energy at the moment they collide with something.

A concise mathematical statement would require more detail, but these generalizations are sufficient for our purposes.

In 1865, some sixty-two years after Young performed his double-slit experiment, James Clerk Maxwell developed a set of equations describing how light propagated as an electromagnetic wave. The equations were so profound that when they were experimentally verified, they left no doubt that light was, in fact, a wave. Even so, just four years later, Johann Wilhelm Hittorf started a chain of events that would once again call the wave nature of light into question. Hittorf observed that when two conducting plates were placed at opposite ends of an evacuated glass tube, and the plates were hooked to the two ends of a battery, the end of the tube connected to the battery's positive side would glow.

When appropriate fluorescent materials are introduced into such a tube, a ray of light called a cathode ray appears stretching from the negative emitter plate to the positive receptor plate. In 1897, J. J. Thomson was able to demonstrate that cathode rays were actually composed of small charged particles that we now know as electrons.

In an ironic twist, Heinrich Hertz, while performing experiments validating the existence of Maxwell's electromagnetic waves, made an accidental discovery. Hertz noticed that if light was shone on what amounted to the emitter plate in a cathode-ray tube, it increased cathode-ray activity. This in itself wasn't altogether puzzling since light waves carry energy. Hertz noted the result and left things at that. Years later, Hertz's student, Philipp Lenard, would perform follow-up experiments demonstrating that the activity couldn't be adequately explained by the way waves transmit energy a little at a time. The result remained an enigma until 1905, when Einstein, building on Thomson's work, showed that if light energy arrived in discrete packets, it would explain Lenard's experimental observations. It was as if light particles were bumping into electrons like billiard balls striking billiard balls. It was for this work "in particular"[5] that Einstein received his 1921 Nobel Prize.

Faced with seemingly contradictory evidence, physicists reluctantly accepted the fact that light sometimes behaves like a wave and sometimes like a particle. Today, most physicists refrain from saying that light is both a wave and a particle in recognition of the fact that, as far as we comprehend waves and particles, these are contradictory positions.

Things only got stranger. In his 1924 doctoral dissertation, Louis de Broglie had the audacity to suggest that light wasn't alone in leading a dual

[5] Svante August Arrhenius, "Award Ceremony Speech for Albert Einstein" (Nobel Prize award ceremony, Stockholm, Sweden, December 10, 1922). The award presentation speech called out this aspect of Einstein's contributions "in particular" but also recognized Einstein's other contributions. Of interest is the fact that, while the committee made mention of relativity theory, it sidestepped explicitly including relativity as a basis for Einstein receiving the award, as at the time relativity remained unaccepted in some scientific circles.

life. Physicists had come to accept light as a wave and only later accepted its particle nature. The young de Broglie wondered if particles—like electrons—also behaved like waves. They do. In the award ceremony speech for de Broglie's 1929 Nobel Prize, the committee chairman laid out the significance of de Broglie's work against the backdrop of the past.

> The 19th century sealed the victory of the wave theory [of light over the corpuscular (particle) theory of Newton]...The 19th century was also the period when atomic concepts have taken root into physics.
>
> Under the influence of these two currents of ideas the concept which 19th century physics had of the universe was the following. The universe was divided into two smaller worlds. One was the world of light, of waves; the other was the world of matter, of atoms and electrons. The perceptible appearance of the universe was conditioned by the interaction of these two worlds.
>
> Our century taught us that besides the innumerable light phenomena which testify to the truth of the wave theory, there are others which testify no less decisively to the correctness of the corpuscular theory...It thus seems that light is at once a wave motion and a stream of corpuscles. Some of its properties are explained by the former supposition, others by the second. Both must be true.
>
> Louis de Broglie had the boldness to maintain that not all the properties of matter can be explained by the theory that it consists of corpuscles...It is [now an experimentally verified] fact that matter has properties which can be interpreted only by assuming that matter is of a wave nature...
>
> Hence there are not two worlds, one of light and waves, one of matter and corpuscles. There is only a single universe. Some of its properties can be accounted for by the wave theory, others by the corpuscular theory.[6]

---

[6]Carl Wilhelm Oseen, "Award Ceremony Speech for Prince Louis-Victor Pierre

The original experimental verification of de Broglie's hypothesis came from the examination of electrons. The double-slit experiment wasn't initially used to demonstrate electron interference patterns since there were other experiments better suited for charged electrons. But the storied history of the double-slit experiment has led researchers to perform it on electrons over the years, and the telltale interference pattern can be seen just as clearly as with light. More recently, far more massive particles consisting of thousands of protons, neutrons and electrons have exhibited the same wavelike characteristics.[7] At what point does it stop? Does it stop? Given the right experimental setup would baseballs show wavelike interference patterns?

The double-slit experiment was also at the center of a pedagogical experiment proposed by Richard Feynman in the early 1960s—an experiment with such a strange outcome that Feynman believed it represented the epitome of quantum weirdness.

The setup of Feynman's experiment was identical to that of the double-slit experiment as it was described for light. But rather than shooting a continuous beam of electrons toward the slits, Feynman proposed shooting electrons one at a time. This would eliminate the possibility that perhaps beams of electrons somehow bounced around to create an interference pattern (though this was already grasping at straws). Surely individual electrons that made their way through the double-slit apparatus couldn't interfere with themselves.

At one level, they don't. If we shoot individual electrons randomly toward the double slits, those that make it through arrive at the detection device as particles. They strike the detection plate at a specific location that can be recorded. Without any doubt, the detected particles are electrons that have been shot by the electron gun.

Raymond de Broglie" (Nobel Prize award ceremony, Stockholm, Sweden, December 10, 1929).

[7]Sandra Eibenberger et al., "Matter-Wave Interference with Particles Selected from a Molecular Library with Masses Exceeding 10000 AMU," *Physical Chemistry Chemical Physics* 59 (2013): 14,696.

The electrons don't behave like particles in that they don't form a pattern that is line-of-sight with the electron gun; see the left half of Figure 6-7. Instead, as more and more electrons show up at the detector, dot by dot they form a wave-interference pattern as in the right of this figure. No reasonable explanations exist for this phenomenon in terms of particles as we know them. But there's a beguilingly simple mathematical explanation in terms of waves.

Any wave can be described by a corresponding wave function that gives the height of the wave at any given point at any given time. It turns out that we can find a wave describing the interference pattern built up by the individual electrons. This remarkable function has the property that wave intensity corresponds to the probability that an individual electron will arrive at a particular location.

Imagine, for example, two regions X and Y on our electron detection plate on the right half of Figure 6-7, and suppose we calculate from our wave function that the average wave intensity reaching X is, say, twice as large as the average wave intensity reaching Y. Then the probability that an individual electron that makes it through one of the slits will land in region X is twice as large as the probability it will land in region Y. Thus, according to the laws of probability, when we shoot enough electrons, the pattern they form on the detector will begin to look like the interference pattern predicted by the wave.

This correspondence between wave intensity and particle probability provides a solid mathematical description that's been experimentally verified time and again. But it raises a deeply unsettling question. The electrons leave the electron gun as particles and arrive at the detection device as particles. But the points where they arrive aren't line-of-site and are best described by our understanding of waves. How do the electrons actually travel from the gun to the detector?

It's a question that has long plagued physicists and philosophers. At the fifth Solvay conference held in 1927 in Brussels, twenty-nine attendees met to discuss the contradictions posed by the apparent wave/particle nature of matter. Seventeen of the attendees had won or would win Nobel Prizes.

The conference was lively to say the least. Many of the attendees weighed in, but the two for whom the conference is best remembered were Albert Einstein and Niels Bohr. Both men were giants in physics, Bohr having established a model of the atom in which electrons surround a centralized nucleus in quantized energy levels. But when it came to quantum theory, Bohr and Einstein represented two different camps.

In what later came to be known as the Copenhagen interpretation, the Dane Bohr headed down a path that called into question physical reality. A particle leaves the electron gun, he argued, but remains in an indeterminate state until it arrives at the detection device where, upon being detected, it reveals itself once again to be a particle.

There are a variety of ways to put words to the mathematics underlying Bohr's position. We could say the electron turned from a particle into a wave and remained a wave until the act of observation caused it to once again turn into a particle. Alternatively, we could say that the route by which the particle traveled is simply irrelevant. We don't care how it got there. We see where it is when we look for it, we know the likelihood of where we can expect to find it, and that's all that matters. Or we could say that the electron existed in superposition, in all possible states, until the act of observation caused the wave function to collapse into one particular state—the state we observe. This latter description is the most commonly used today.

Whatever words we choose, it's clear that Bohr was abandoning any commonsense interpretation of particle movement. When a particle moves from point A to point B, it may take any number of routes. But it is a particle, and so it must follow some route, mustn't it? Bohr said no.

Einstein wasn't ready to accept such an outlandish position. He firmly believed that physics was in the business of understanding reality. Particles were particles. Waves were waves. If particles appeared to behave like waves, or vice versa, there had to be an explanation.

Even more, Einstein wasn't comfortable with the nondeterminism associated with Bohr's claims. Newton gave us a very orderly universe. If I throw two rocks into the air at identical speeds in identical directions, they will land on the ground at precisely the same point. Rocks and everything else in the universe are bound by the laws of nature.

Under this view, if we shoot an electron from a gun—even allowing for the fact that because it's so small, we may have some trouble determining its exact speed and direction—we should have a pretty good idea of where it will go. But we don't. According to the Copenhagen interpretation, all we know is the likelihood that it will arrive at any one of an infinite number of possible locations. Einstein's work on relativity had been groundbreaking—its results counterintuitive—but it never called into question the deterministic universe of Newton. It was in this sense that Einstein uttered his famous phrase "God does not play dice with the universe." He was wed to determinism.

Feynman went on to describe one further experiment, one that only bolstered the Copenhagen interpretation. Feynman imagined an experiment in which we look to see which slit each electron passes through. Sure enough, if detectors are set up near the slits, we observe roughly an equal number of individual electrons passing through each. However, the double-slit interference pattern on the right half of Figure 6-7 goes away, replaced by two single-slit diffraction patterns similar to the one found on the right half of Figure 6-4. It seems that when we look for electrons at the slits, we see electrons, after which they continue to the detection device as if there had only been the single slit through which they passed.

The fact that such behavior is inexplicable in any way that coincides with our understanding of particles and waves is what kept Einstein from accepting Bohr's position. Bohr was willing to let go of physical reality, of how the world *is*, content to accept a theory solely on its ability to make accurate predictions. "It is wrong to think the task of physics is to find out how nature is," said Bohr. "What we call science," Einstein argued, "has the sole purpose of determining what is."[8]

Bohr and Einstein continued their spirited debates—in print and face to face—until Einstein's death some three decades later. The debates were gentlemanly. Bohr and Einstein were friends and had great respect for one another. But they were also stubborn. And who can blame them? At stake was not just the fundamental nature of reality but how the scientific endeavor should proceed.

Over the intervening years, this latter question has been resolved in Bohr's favor—sort of. Scientists deal with wave/particle duality as a matter of course, simply because it makes accurate predictions. Thus science marches on. But do scientists accept Bohr's metaphysical interpretation? Some do, some don't. Many others simply ignore the philosophical implications altogether, turning to a well-known pragmatic interpretation: "Shut up and calculate."[9]

Others continue to pursue different means of explaining the unexplainable. In the mid-1950s, a graduate student at Princeton by the name of Hugh Everett put forth a radical interpretation of wave/particle duality in his doctoral thesis. Rather than admit the nondeterminism inherent in the Copenhagen interpretation, Everett proposed that the superposition of states embodied in the wave function didn't collapse into a single state when we went looking for an electron. So how is it that we observe the electron striking only one location on the detector if it's arriving at all possible locations? Because that's where we observe it in *our* universe. In Everett's interpretation, the "universe" is constantly spawning new parallel universes, each corresponding to a different outcome.

The advantage of Everett's "many-worlds" interpretation is that it does away with the role of the observer. The Copenhagen interpretation

---

[8]Manjit Kumar, *Quantum: Einstein, Bohr, and the Great Debate About the Nature of Reality* (New York: W. W. Norton, 2008), 262.
[9]Variously attributed to David Mermin and Richard Feynman.

puts forward the principle that the wave function collapses into a particle due to the act of observation. How can it be that an act of observation brings about physical reality? It grants to the observer godlike qualities and, in the process, opens its own set of philosophical questions.

The many-worlds interpretation also does away with the nondeterminism that so troubled Einstein. Particles don't randomly arrive at one point or another. They arrive at *all* points. Once again, we don't observe all the different outcomes; we only see the outcome that occurred in our particular universe. The mathematics works so cleanly that many eminent physicists embrace some form of the many-worlds interpretation.[10]

Of course, the disadvantages are clear. The many-worlds interpretation posits the existence of a staggeringly large number of parallel universes, only one of which we interact with (although some of these universes have individuals very much like ourselves, spare a particle or two). Further, the number of parallel universes continues to grow forever, branch by branch. And this is not just the result of a single electron in a single experiment, but in the ongoing interaction of all particles of all shapes and sizes—including, as was pointed out in de Broglie's award ceremony speech, "ourselves."[11] We, too, are very large particles.

Whatever way we choose to interpret the results of the double-slit experiment, we're left with nagging, unanswered questions. The double-slit experiment hits us in the face with repeatable, undeniable experimental observations. We see what we see, but what we see doesn't correspond to any known reality. And as we probe more deeply to find an explanation that fits within our mental apparatus, we're forced to propose seemingly absurd answers. The double-slit experiment leaves us staring in astonishment at a universe that's truly more amazing than we can hope to comprehend.

---

[10]Together with the anthropic principle, as discussed in the final chapter, the many-worlds interpretation has also been used to explain how intelligent life appeared in our universe.

[11]Oseen.

References and Further Reading

1. Arrhenius, Svante August. "Award Ceremony Speech for Albert Einstein." Nobel Prize award ceremony, Stockholm, Sweden, December 10, 1922. www.nobelprize.org/nobel_prizes/physics/laureates/1921/press.html. Accessed July 11, 2017.

2. Byrne, Peter. *The Many Worlds of Hugh Everett III: Multiple Universes, Mutual Assured Destruction, and the Meltdown of a Nuclear Family.* New York: Oxford University Press, 2010.

3. Eibenberger, Sandra, Stefan Gerlich, Markus Arndt, Marcel Mayor, and Jens Tüxen. "Matter-Wave Interference with Particles Selected from a Molecular Library with Masses Exceeding 10000 AMU." *Physical Chemistry Chemical Physics* 59 (2013): 14,696–14,700. See also the Royal Society of Chemistry, pubs.rsc.org/en/content/articlehtml/2013/cp/c3cp51500a. Accessed July 11, 2017.

4. Feynman, Richard. *Feynman Lectures on Physics, Volume III: Quantum Mechanics.* New York: Basic Books, 2011. See also the California Institute of Technology, www.feynmanlectures.caltech.edu. Accessed July 11, 2017.

5. Kumar, Manjit. *Quantum: Einstein, Bohr, and the Great Debate About the Nature of Reality.* New York: W. W. Norton, 2008.

6. Norton, John. D. "Einstein's Miraculous Argument of 1905: The Thermodynamic Grounding of Light Quanta." Prepared for HQ1: Conference on the History of Quantum Theory, Max Planck Institute for the History of Science, July 26, 2007. See also the PhilSci archive, philsci-archive.pitt.edu/3437. Accessed July 11, 2017.

7. Oseen, Carl Wilhelm. "Award Ceremony Speech for Prince Louis-Victor Pierre Raymond de Broglie." Nobel Prize award

ceremony, Stockholm, Sweden, December 10, 1929. www. nobelprize.org/nobel_prizes/physics/laureates/1929/ press.html. Accessed July 11, 2017.

8. Robinson, Andrew. *The Last Man Who Knew Everything: Thomas Young, the Anonymous Polymath Who Proved Newton Wrong, Explained How We See, Cured the Sick, and Deciphered the Rosetta Stone, among Other Feats of Genius.* London: Pi Press, 2005.

9. Rosenblum, Bruce, and Fred Kuttner. *Quantum Enigma: Physics Encounters Consciousness.* New York: Oxford University Press, 2011.

10. Young, Thomas. "The Bakerian Lecture: On the Theory of Light and Colors." *Philosophical Transactions of the Royal Society of London* 92 (1802): 12-48. rstl.royalsocietypublishing.org/content/92/12.full.pdf+html. Accessed July 11, 2017.

Faith is a dark night for man, but in this very way it gives him light.

—John of the Cross

Chapter Seven
Dark Matter

Among nature's most beautiful displays are galaxies. Consisting of billions or trillions of stars, they're as unimaginable as they are alluring. Gravity brings these stars together in the same way gravity from the sun lassos the planets in our solar system. And like the planets orbiting the sun, the stars in most galaxies revolve around a common center.

In the early 1960s, astronomers Vera Rubin and W. Kent Ford turned their attention to our nearest neighbor, the Andromeda galaxy. At question was the speed with which the spiral galaxy was rotating about its center. More specifically, they were interested in the speed of stars at various distances from the galactic center. The work wasn't overly exciting, but it provided an environment conducive to Rubin and Ford's lifestyles, and it made good use of a new observational technique developed by Ford—a technique that allowed spectral emissions to be recorded much more quickly than had been done before.

Spectral emissions are important in identifying the speed with which a distant object is moving. In the same way the siren of an emergency vehicle sounds different when the vehicle is moving toward or away from someone, so, too, do light waves change color depending on the motion of a celestial object. This color change is captured in the spectral emissions of that object.

Rubin and Ford weren't looking for surprises, but what they saw helped propel science in a strange new direction that remains at the forefront of cosmological research to this day. To understand why Rubin and Ford's observations were so mystifying, recall two basic facts we know about gravity.

1. The more massive an object is, the more gravitational force it exerts.
2. The further we move away from an object, the less gravitational force it exerts.

Consider what this means for a planet traveling around the sun in a circular orbit at a distance r.[1] Given this value, we can calculate the velocity v at which the planet must be moving in order to stay in this orbit. That velocity is:

$$v = square\_root(GM/r)$$

where M is the mass of the sun, G is the universal constant of gravitation, and it's assumed that the mass of the orbiting planet is much smaller than that of the sun. Of particular interest is how the velocity changes as the distance from the sun increases: as the distance r goes up, the velocity v goes down. Here we see the key point: planets farther from the sun move more slowly than those closer in. This makes intuitive sense by virtue of the fact that gravity is weaker the farther we move away from the sun. If the orbital velocity of the earth were somehow magically transferred to a planet farther away from the sun, gravity wouldn't be strong enough to hold the planet in a circular orbit. Likewise, if the earth's velocity was transferred to a planet closer to the sun, the stronger

---

[1]Planets can and do orbit in elliptical orbits that are nearly circular. The assumption of circularity is sufficient for our purposes and greatly simplifies the discussion.

gravity would cause that planet to get pulled closer to the sun. The velocity required to keep a planet in a circular orbit around the sun is plotted as a function of its distance from the sun in Figure 2-1.

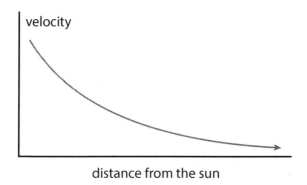

Figure 2-1: The velocity required for a planet to remain in a circular orbit decreases the farther the planet is from the sun

The situation for the stars in the Andromeda galaxy studied by Rubin and Ford is quite similar to that of the planets in our solar system. Andromeda is a spiral galaxy, meaning it's relatively flat with stars revolving around a common center of gravity. There is, however, a significant difference between the movement of the planets in the solar system and the movement of stars in the Andromeda galaxy. In the solar system, the sun is far more massive than any of the planets. As such, the sun *is* the center of gravity.[2] In the Andromeda galaxy, the center of gravity is defined by contributions of the individual stars taken together. An example helps clarify the situation.

Consider the imaginary galaxy consisting of three equally massive stars located at three points defined by an equilateral triangle, as depicted in Figure 2-2. What would be the effect of gravity at the center of the triangle? Since the mass of each star is the same, and since the center of the triangle is of equal distance to each star, the gravitational pull balances out. The net force experienced by an object placed in the center of the triangle would be zero, and as a result it wouldn't move.

[2]At least to a very good approximation.

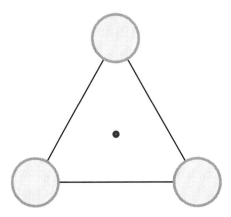

Figure 2-2: Three equally massive stars situated at the
three vertices of an equilateral triangle

The stars themselves, however, aren't so fortunate. Due to the
gravitational pull of the other two stars, each star would find itself
drawn toward the center of the triangle, where it would collide with
the others.

This situation can be avoided if the stars are traveling in a common
circular orbit at a precise velocity around their center of gravity, just as
the planets move around their center of gravity, the sun. Although the
mathematics is more daunting for galaxies given the number of stars
they contain, the basic principle remains the same: the stars revolve
around a common center of gravity.[3]

We're now in a position to consider how the force of gravity changes
as we move away from a spiral galaxy's center of gravity into deep space.
By definition, the net force an object experiences when placed at the
center of gravity is zero. As shown in Figure 2-3, as an object moves
away from the center of gravity along the depicted line, the mass to the
left of the object increases while the mass to the right decreases. The
net result is that the object begins to feel the net force of gravity pulling

[3]The dynamics are such that the center of gravity moves, and the stars don't
travel in absolutely circular paths around it. Here again, however, the assumption of circularity around a relatively stationary center of gravity is sufficient
for our purposes.

Figure 2-3: The force experienced by an object increases then decreases as it moves away from the galactic center of gravity

it back to the left. At first the tug is gentle, but as it moves farther and farther away from the center of gravity, it experiences more force from the stars to its left and less from those to its right.

The increasing force can't, however, continue forever. At some point, most of the stars are situated to the left of the object, and as it moves farther to the right the object is simply moving away from all of the sources of gravity. Thus, somewhere in the outer reaches of the galaxy, the force on the object reaches a maximum and begins to decrease, again as depicted in Figure 2-3.

As with the planets revolving around the sun, the velocity of a star revolving around its galaxy's center of gravity is related to the gravitational pull it feels toward that center: the greater the gravitational force, the faster the star must move to stay in its orbit. Thus, if we were to measure the rotational velocity of stars in the Andromeda galaxy, we'd expect it to at first get larger as we moved outward from the galactic center, but then begin to decrease again, as shown in Figure 2-4.

In gathering together their results, Rubin and Ford fully expected to see a curve that looked like that shown in Figure 2-4. Instead, what they saw was a curve that looked like that shown in Figure 2-5. The result was mystifying. Stars in the outer reaches of the galaxy were

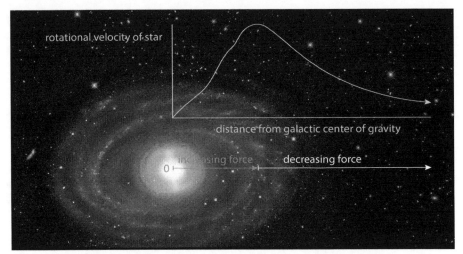

rotational velocity of star

distance from galactic center of gravity

increasing force   decreasing force

0

Figure 2-4: The predicted rotational velocity of stars at
varying distances from the galactic center of gravity

traveling too fast. They should have been flying out of their orbits and
into the outer reaches of space. That is, unless there was more matter
than met the eye.

Consider the following game played by astronomers. Given additional
matter—as much as you like—can you position it within a galaxy in such
a way that it makes stars move at a velocity consistent with the results
observed by Rubin and Ford? Using computer simulations, astronomers
demonstrated that the answer is yes, but it required a lot of additional
matter—far more than what could be accounted for with visible matter
alone. Perhaps, astronomers reasoned, additional matter was out there,
but it was dark and thus unobservable.

The idea of dark matter wasn't new; it dated from at least the
beginning of the twentieth century. One important example among
many can be found in the 1933 work of Fritz Zwicky, based on observations of a collection of galaxies known as the Coma Cluster. Using
a formula taken from the field of thermodynamics, Zwicky estimated
that the galaxies in the cluster were moving relative to one another at
speeds over ten times faster than predicted by estimates of observed
mass. "If this would be confirmed," he wrote, "we would get the sur-

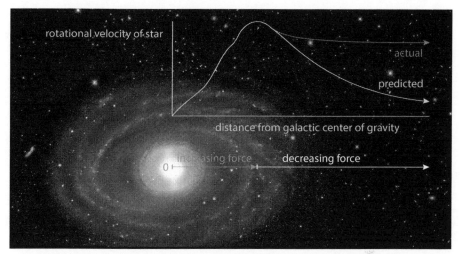

Figure 2-5: The predicted and actual rotational velocity of stars at varying distances from the galactic center of gravity

prising result that dark matter is present in much greater amount than luminous matter."[4]

As evidence mounted over the decades, Zwicky's "surprising result" seemed more and more likely. But what kind of dark matter? We tend to think of space as primarily filled with luminous stars. Yet, even at the beginning of the twentieth century, astronomers were well aware that space contained entities that don't shine so brightly. Planets. Clouds of gas particles. Dead stars. It wasn't hard to imagine all sorts of phenomena that couldn't be observed, especially over long distances. The presumption was that dark matter wasn't anything special—it just wasn't very bright.

That would change, however, in the decades that followed Rubin and Ford's observations. Whereas Rubin and Ford had almost certainly conceived of dark matter as typical, day-to-day *baryonic* matter made of protons, neutrons and the like, by the late 1980s, the leading scientific theories had turned in favor of some yet-to-be-discovered exotic matter. The reasons were many and not easily condensed into a simple story line. However, certain facts stand out.

[4]Stephanie M. Bucklin, "A History of Dark Matter," *Ars Technica*, accessed April 16, 2017, arstechnica.com/science/2017/02/a-history-of-dark-matter.

Models of how the universe formed and observational data provided strong evidence that baryonic matter didn't exist in the quantity needed to fill the gravitational gap. Looking for other options, scientists landed on neutrinos. Neutrinos are small even by the measure of subatomic particles, and they're difficult to detect because, among other things, they don't have an electric charge like their electron cousins. As a result, experimental detection is very difficult, making neutrinos all but invisible. They do, however, interact with other matter via gravitational attraction.

It would seem that neutrinos would be good candidates for dark matter. Unfortunately, these fast-moving particles failed to hold up to scrutiny and were ruled out. Nonetheless, they'd set the stage in the search for other similar particles to fill in the missing gravitational mass. Various theories of this nature were put forward, all of which were captured under the title of Weakly Interacting Massive Particles, or WIMPs for short.

WIMPs aren't the only game in town. Other theories remain in contention. Theories that continue to focus on dark baryonic matter go under the title Massive Astrophysical Compact Halo Objects theories, or MACHOs, though there's little evidence enough MACHO matter exists to fill the gravitational gap. And some researchers are pursuing modifications to the laws of gravity as a means to explain astronomical observations, though this is challenging since any modifications would have to fit into the large, well-established framework provided by the present laws of gravity. As a result of problems with the alternatives, WIMPs are leading the pack. We're willing to at least provisionally accept that particles we don't understand and can't detect—apart from presumed gravitational attraction—are at work holding galaxies together.

Astrophysicists are busily attempting to better explain what these particles are beyond the fact that they exert gravity. It would be a major breakthrough to develop an understanding sufficient to experimentally verify their existence by some other means. Simply inferring their existence to make the current theory of gravity work raises all sorts of questions. If we can't say anything more about these particles, how well can we say we comprehend them, if at all?

References and Further Reading

1. Bertone, Gianfranco, and Dan Hooper. "A History of Dark Matter." Cornell University Library archive, arXiv.org. arxiv.org/abs/1605.04909. Accessed July 11, 2017.
2. Bucklin, Stephanie M. "A History of Dark Matter." *Ars Technica.* arstechnica.com/science/2017/02/a-history-of-dark-matter. Accessed July 11, 2017.
3. Panek, Richard. *The 4% Universe: Dark Matter, Dark Energy, and the Race to Discover the Rest of Reality.* New York: Houghton Mifflin Harcourt, 2011.

What can be said at all can be said clearly, and what we cannot talk about we must pass over in silence.

—Ludwig Wittgenstein

Chapter Eight
Reflections

The universe as we know it runs according to well-defined laws. If I throw a rock into the air with a given speed and direction so that it lands thirty feet in front of me, then do so again, the rock will again land thirty feet in front of me. That's the very notion of a natural law. But why, we might ask, thirty feet? Why doesn't the force I apply cause the rock to land twenty feet away? Or forty feet?

The rock lands where it does because of certain constants of nature—numbers underlying the equations that describe how gravity works. If we could change any of these constants while leaving the underlying laws alone, we would, in fact, see that the rock travels more or less than thirty feet following a path of similar shape. Why are the constants the values they are? We don't know. An explanation based on a deeper understanding of our physical universe would be very exciting, but it's not clear that such an explanation exists. For now, and possibly forever, we must content ourselves with simply measuring these important constants.

We can, however, ask the following question. If we were to change the values of any constants, would it significantly change how the universe works? An analogy is that of turning up a car radio. As we begin to increase the volume the music just gets louder, but at a certain point we blow out the speakers.

Physicists would typically expect that, like the volume on the radio, we can change these constants over a reasonable range without the universe acting radically differently. This in turn might lead us to ask the following intriguing question: how much can we change the constants of nature and still have a universe capable of supporting human life? If we dial the constants up or down, would human life still come about? Or if we so much as breathe on the dial, is it lights out?

It turns out that if we change some of the constants by only a hair's breadth, the universe becomes inhospitable to human life. In fact, a hair's breadth is a vast exaggeration. In the case of the cosmological constant, which influences the rate at which the universe is expanding, we can't move the dial by more than $1/10^{120}$. That's a frighteningly small number given that 10^{120} is trillions of trillions of trillions times bigger than the number of atoms in the known universe. And it raises a perplexing question: how is it that the big bang gave rise to such an exceedingly unlikely universe—a universe balanced in just the right way to support human life?

It's a question that's left physicists, philosophers, and theologians scratching their heads. Yet this seemingly impossible state of affairs can be resolved with a surprisingly simple argument. Consider the following syllogism.

> If the constants of nature were such that the universe couldn't support human life, then human life wouldn't exist.
>
> Human life exists.
>
> Therefore, the constants of nature are such that the universe can support human life.

The argument, a version of what's known as the *anthropic principle*, stands on its head the question about the precise balance of the constants of nature. Because we're here to ask the question, it shouldn't be surprising that the constants of nature support human life. If the universe didn't support human life, we wouldn't be here to ask it. We ask, therefore it is.[1]

The beauty of the anthropic principle isn't the depth of reasoning that lies behind it. Quite the contrary. It's been criticized as a tautology—a simple statement of fact leading nowhere. Okay, say its critics. We accept your argument. But it doesn't tell us anything about why the constants of nature are what they are. Isn't it still amazing that against all reasonable odds, we live in a finely tuned universe, a universe tuned in such a way as to support human life? To which proponents of the anthropic principle simply smile and shrug their shoulders.

The Beyond Comprehension (BC) conjecture—that there are real things in our world that humans can't comprehend—shares many similarities with the anthropic principle. Like the anthropic principle, most people have never thought about it. When they do, it seems both a bit eye-opening and self-evident at the same time, though it's not tautological.[2]

Another similarity is the propensity to dismiss the BC conjecture as interesting but nonetheless irrelevant. Things that are beyond comprehension are *beyond comprehension*. They're outside the realm of our minds' capabilities. Asking humans to do anything relevant in domains we can't grasp is like asking a dog to make sense of arithmetic. It may be fun to ponder, but we live in the world we perceive.

---

[1]Colleague Sarah Fishman made a connection between the anthropic principle and a game many children play when they ask the question, "What would I be like if my parents had never met?" Of course, it's a meaningless question, since if your parents hadn't met, the "you" to whom you're referring wouldn't be around to ask the question.

[2]There's nothing logically contradictory in arguing that the human mind is capable of understanding everything real, and the word "real" carries its own set of baggage.

Yet it's here that the similarity with the anthropic principle shines through: neither argument goes quietly. They're nagging and even a bit annoying, like the spouse who says I told you so. He or she may be right, but does it help to be reminded of the fact? In the case of the limitations of the human mind, most scholars would be hard-pressed to argue that the three pound chunks of organic matter we call our brains are up to the challenge of knowing everything. The BC conjecture isn't only tenable, it seems scientifically all but incontrovertible. But accepting the BC conjecture in practice is another matter. Doing so requires assessing whether an argument, however logically constructed, is built on a foundation that doesn't take into account some unknown realities. Who's to say when an argument is crossing into territory where it has no right to set foot?

Consider, for example, two commonly proposed resolutions of how the constants of nature came to be so finely tuned as to support human life. One is argument by design. An intelligent designer put the universe together in such a way that human life would spring up. Another is argument by chance. There exists a multiverse of many universes similar to ours growing out of their own big bangs. As time passes, each develops with its own different constants of nature. If we assume enough such universes, then one or more will evolve in such a way as to support human life. We just happen to live in one that does.

So what of these proposed resolutions? The first requires the existence of a designer of immense power—enough power to bring into existence not just the universe but a universe tuned to admit human life. The second relies on the existence of a vast, quite probably infinite collection of universes, only one of which we can observe.[3] Is either correct? Or is it possible we're simply in over our heads?

Another example surrounds claims regarding conscious life. Setting aside the problematic question of determining if a machine *is* conscious, we can ask if it's possible for a machine to *become* con-

---

[3]There are, of course, many variations on these themes.

scious. Is it possible for a machine to feel love or want or pain or any of the other personal, internal sensations we associate with being conscious? At present, we can only point to examples of conscious biological life. But if brains are thought of as computers, and if the chemicals and electrical signals that move through living brains are thought of as computer programs, why should consciousness be limited to biological life? If the programs found in living brains are replicated on sufficiently powerful machine-computers, many would argue, it's only logical that these programs-on-machines would develop consciousness. Others argue that there's something unique to biological organisms that allows consciousness to spring forth, though exactly what that unique feature is they can't point to. Still others argue that the whole concept of consciousness is misguided and that consciousness doesn't exist, at least not in the way we think it does. Are any of these positions correct, or is the existence of consciousness something we can't consciously comprehend?

Still another example can be found in disagreements over the nature of free will. Are humans capable of making decisions of their own free will? Or are our actions deterministic—merely the result of the laws of nature being played out within our bodies? A common metaphor for determinism is that of balls on a billiard table. If I strike the cue ball, it in turn causes other balls on the table to move. If we know where the balls started and know the initial force and direction imparted to the cue ball, we can precisely predict where the balls will be in the future. There are no decisions for the balls to make. The laws of nature direct the balls on their deterministic paths.

The human body, goes the argument for determinism, is composed of atoms and molecules interacting according to the laws of nature in the same way as billiard balls. We therefore have no free will to make decisions—at least not in the sense that most people understand free will. We may think we have free will, but our bodies are simply acting out a deterministic sequence of atomic motions. Some of these motions lead us to mental states where we feel we've made a decision, but that's an illusion. The entire universe, once set in motion at the time of the big bang, has been striding deterministically forward, acting out motions dictated by the laws of nature.

The argument can get much more convoluted. For example, the apparent randomness inherent in some interpretations of quantum theory suggest that the world may not be entirely deterministic. But randomness doesn't come to the rescue of free will. Free will doesn't posit random decisions; it posits reason and action stemming from some unknown qualities of the human brain that somehow elude the laws of nature, at least as we comprehend them.

So what of the argument for determinism? On the surface, it's compelling, fitting neatly into the scientific framework we've established over the course of human history. But it's also positing solutions to the nature of something we don't fully grasp to begin with: consciousness. We still don't know how a small amount of organic matter gives rise to inner thoughts, sensations, and feelings—to mind/brains capable of remarkable feats of creativity. Is determinism self-evident, or is it an unwarranted extrapolation beyond the limits of our comprehension?

It's hard to know when we've overstepped our cognitive limitations when making an argument. In fact, we typically can't—at least not with any certainty. But we should nonetheless remain aware of the possibility.

Yet, with rare exceptions, we ignore it. That's not surprising. When seeking to explain something, we rely on the cognitive abilities at our disposal. It's not natural to think an explanation might be beyond our means. On occasion we confront situations that defy explanation, like Torricelli's trumpet, wave/particle duality, and dark matter. But in other instances, the answer's not as clear.

While it's impossible to lay out hard-and-fast rules establishing what's beyond comprehension, there exist some indicators that signal when we may be in over our heads.

1. *Concepts we consider unequivocally true are shown false.* The arguments in this book draw upon examples of this nature to make a case for the BC conjecture. A container with finite volume can't have infinite surface area in any way we conceive of "volume" and "surface area." Yet we can point to many such containers. It could be argued that this indicator goes so far as to constitute proof that something's beyond comprehension. However, such a claim forces a further examination of "concepts we consider unequivocally true." We therefore content ourselves to take this as a strong indicator rather than a proof.

2. *Infinity is involved.* From Hilbert's grand hotel to Cantor's transfinite numbers, infinity teases us, even if it hasn't altogether eluded us. Calculus, with its focus on the infinitesimally small, is one of the great achievements of the human mind. It tames infinity, providing a firm foundation on which to build the mathematical edifice underlying so much of our science and technology. But it achieves this by sidestepping the notion of infinity, and our very efforts to sidestep the issue open the door for all sorts of mathematical monsters. Taming infinity is different from comprehending infinity. When an argument relies on the physical existence of something infinitely big or small, we need to be wary.

3. *Experimental verification doesn't exist.* The universe, as we've discovered, operates according to mathematical laws, which in most instances are strikingly simple. As such, we've developed a sense of trust that an argument based on an elegant mathematical solution is likely to be true by virtue of the fact that it's an elegant mathematical solution. Even if there's no way

to verify the argument through experimental observation, many scientists believe elegant math by itself has become a compelling argument in favor of truth.

A good example can be found in arguments postulating the existence of multiple universes. Coupled with the anthropic principle, these arguments provide a logical explanation of how our universe might have come to support human life. Yet most proponents of multiverse theories admit we'll never be able to verify the existence of universes other than our own, thus abandoning the keystone of the scientific method: repeatable experimental validation. But in this case, proponents argue, if other universes exist, their existence answers a lot of questions—to which an advocate of the BC conjecture would add, *assuming your mind/brain is properly forming these questions to begin with.* Is it better to embrace the existence of something we can't experimentally verify, or to consider the possibility we're dealing with something beyond comprehension?

*****

As a child I was suspicious of meat loaf and refused to eat it. That changed abruptly thanks to the scheming of my mother. When, years later, I shared with my wife the details of how I came to like meat loaf, she laughed and shared a story about the traditional Jewish dish, kreplach.

A young boy, goes the story, wouldn't eat kreplach. So his mother took him into the kitchen to show how she made it.

"See this noodle?" asked the mother.

"Yes," replied the boy.

"You like noodles, yes?"

"I like them very much."

The mother then took a piece of meat and showed it to the boy.

"See this meat?" asked the mother.

"Yes."

"You like meat, yes?"

"I like it a lot."

The mother then proceeded to wrap the meat in the noodle.

"Ew, yuck!" cried the boy. "Kreplach!"

Unlike the boy in the story, when my mother pointed out the ingredients in her meat loaf—all of which I liked individually—I decided I hadn't given it a fair shake.

Like kreplach, some find the BC conjecture difficult to swallow. The ingredients go down easily enough, but the dish as a whole causes intestinal discomfort. Consider the following position a critic of the BC conjecture might take, and a response that might be offered.

Critic: I accept the conjecture that the human brain, and thus the human mind, has cognitive limitations. The brain is pretty amazing, but it's finite and subject to the laws of nature just like all other entities in our physical world. Further, as evidenced by the cognitive limitations observed in other animals, it stretches credulity to imagine humans have evolved to a point where we don't have similar, albeit different, limitations. And the examples presented throughout the book provide ample food for thought in support of the conjecture. Be that as it may, the BC conjecture is both useless and potentially dangerous.

It's useless in that it doesn't solve anything. It can't be applied as the basis for an argument that extends our knowledge. Suggesting that something is "beyond comprehension" is simply throwing in the towel. It provides an easy out for giving up.

Even worse, it opens the door to fill the gap with all sorts of unscientific mumbo jumbo. If a problem is beyond comprehension, what stops someone from claiming that ghosts roam the earth? Or that crystals channel vital living forces? It's an excuse to lift pseudoscience into the realm of the respectable, and that's very, very dangerous.

Respondent: I wholeheartedly agree with your concern that the conjecture has the potential to be dangerous. If a claim of something lying beyond comprehension is used to validate an unwarranted belief, it's an abuse of the BC conjecture. The conjecture, however, doesn't hint that it's appropriate for filling in the holes of an unwarranted claim. It simply recognizes that we're likely to run into things we're not wired to deal with.

Does that mean we should throw up our hands in the face of difficult questions? Not in the least. But we should be willing to consider that there may be answers our minds aren't capable of constructing. If we fail to recognize we're not capable of answering a particular question, we might wind up accepting an answer that seems plausible but is wrong. And wrong arguments have the potential to hold back the human endeavor, no matter what their nature.

It's tempting to hypothesize how embracing the BC conjecture might alter the scientific research agenda moving forward. Rather than do so, however, I'd like to conclude with reflections on how the conjecture can impact our lives today.

A look at many of today's scientific and philosophical claims is disconcerting. The claims themselves would strike most people as speculative, yet they're often presented by their purveyors as scientific fact.

Regarding the many-worlds (parallel universes) interpretation of quantum mechanics—that the universe is forever branching into new universes, which themselves branch into even more universes—physicist David Deutsch writes:

The quantum theory of parallel universes is not the problem, it is the solution. It is not some troublesome, optional interpretation emerging from arcane theoretical considerations. It is the explanation—the only one that is tenable—of a remarkable and counter-intuitive reality.[4]

Describing his position on whether it's possible for computers to become conscious, philosopher Nick Boström writes:

A common assumption in the philosophy of mind is that of *substrate-independence*. The idea is that mental states can supervene on any of a broad class of physical substrates. Provided a system implements the right sort of computational structures and processes, it can be associated with conscious experiences. It is not an essential property of consciousness that it is implemented on carbon-based biological neural networks inside a cranium: silicon-based processors inside a computer could in principle do the trick as well.[5]

In an article titled "Are We Really Conscious?," neuroscientist and psychologist Michael Graziano offers a different perspective, calling into question the very notion of consciousness.

I believe a major change in our perspective on consciousness may be necessary, a shift from a credulous and egocentric viewpoint to a skeptical and slightly disconcerting one: namely, that we don't actually have inner feelings in the way most of us think we do.[6]

---

[4]David Deutsch, *The Fabric of Reality: The Science of Parallel Universes—and Its Implications* (New York: Penguin, 1997), 51.
[5]Nick Boström, "Are You Living in a Computer Simulation?" *Philosophical Quarterly* 53, no. 211 (2003): 243.
[6]Michael S. A. Graziano, "Are We Really Conscious?" *New York Times Sunday Review*, October 10, 2014.

Addressing the nature of free will, philosopher and neuroscientist Sam Harris writes:

> Free will *is* an illusion. Our wills are simply not of our own making. Thoughts and intentions emerge from background causes of which we are unaware and over which we exert no conscious control. We do not have the freedom we think we have.[7]

Philosopher Daniel Dennett reaffirms Harris's position, commenting that Harris's writing is

> a fine "antidote"...to this incoherent and socially malignant illusion [called free will]. The incoherence of the illusion has been demonstrated time and again in rather technical work by philosophers (in spite of still finding supporters in the profession), but Harris does a fine job of making this apparently unpalatable fact accessible to lay people.[8]

The individuals making these claims are well read and highly educated, and though there isn't uniform agreement about their claims, many if not most scholars of science and philosophy hold comparable beliefs. It would be possible to fill an entire book with similar quotations.

The important question, however, is if these scholars are correct. When, for example, Deutsch states that parallel universes are the only tenable solution to problems posed by quantum theory, he's arrived at this conclusion through rational evaluation of evidence—evidence *processed through his mind/brain*. Deutsch is willing to take this leap without direct observation of such universes because he can't conceive of any other explanation. But are there explanations of which he can't conceive?

*****

---

[7]Sam Harris, *Free Will* (Washington, DC: Free Press, 2012), 5.
[8]Daniel Dennett, "Reflections on 'Free Will,'" Naturalism.org, 2014, accessed April 17, 2017, naturalism.org/resources/book-reviews/reflections-on-free-will.

The Rothko Chapel sits in a small park located roughly two miles southwest of downtown Houston. To its west lies the building housing the Menil Collection, a building whose simple, elegant lines and trim grounds sit comfortably among the equally simple homes that surround it. To the east sits the campus of the University of Saint Thomas, anchored by the Chapel of Saint Basil, with its own graceful lines and uncluttered interior.

Those who casually visit the Rothko Chapel are usually perplexed. Mark Rothko is one of the twentieth century's most recognized painters, and the chapel and its contents are among his most important works, taking years to complete. Yet to the untrained eye, the result is so austere it's a disappointment.

The building itself is made of sixteen plain, windowless brick walls formed in the shape of a Greek cross. Entry is through a black metal door situated on one of the walls. Upon entering, visitors find an octagonal interior as featureless as the exterior. Monotone tile covers the floor, to which nothing is affixed. A handful of rectangular benches with no sides or backs are scattered in various locations, the arrangement varying from visit to visit based on some unknown algorithm, if any. The only thing that defines the front of the chapel is that it's on the wall opposite the door. Some would dispute that the room has a front.

The most puzzling aspects of the entire experience, however, are Rothko's paintings. All fourteen are black. Slight variations can be seen, though in some cases viewers are left to wonder if they're experiencing an optical illusion—if, in fact, two paintings appear to be a different black, but that it's a result of aging canvases or the angle at which the light's falling on them or the natural variation that occurs when mixing paint.

And the visitor can't help but ask: what does it mean? What did Rothko intend? Art historians can point to the important influences that shaped the artist: to Nietzsche's *Birth of Tragedy* and Matisse's *Red Room*. They can point to Rothko's growing belief in the transcendence of his work and the bouts of depression that ultimately led him to take his life. They can use their knowledge of art to hypothesize what Rothko's intentions were and whether he achieved them.

Yet all works of art ultimately stand on their own—a fact especially true of those by Rothko, who in his later years ceased naming or even discussing his creations. Confronted with the dark meditation he's set before us, we're left to make of it what we can—if we choose to make anything of it at all.

Science and art are vastly different in their aims, and to approach science the way artists approach art would be wrong. Science thrives upon verifiable theories. Art trades in creative expression and the experiential feelings it awakens. Yet artists are willing to embrace something that causes scientists great discomfort: inexplicability. For scientists, the inexplicable cries out for explanation. Artists understand the inexplicable is inescapable.

In our natural drive to understand everything, it's hard to step back and consider that perhaps some things are beyond comprehension. But like art historians viewing the work of a master, if we open ourselves to the possibility, who knows what we may see? At a minimum, we'll walk away a bit more humble for the effort.

As someone who's made a living plying my mathematical skills, I revel in the degree to which we've come to understand the world around us. Physics, chemistry, biology—all animated by our unique grasp of mathematics. And we don't just understand; we use that knowledge to create. From synthetic fabrics to new medicines to computers to everything in between, above, and below, we've forged an astonishing technological landscape. It's impossible to deny what special creatures we humans are—at least among our animal peers in our tiny corner of the cosmos.

Yet it helps to remember that while we're special creatures, we're still just creatures. Whatever else may be true, we can say with absolute certainty that we experience the world as very, very *human* beings. We enjoy a good meal. We get tired when we work. We feel better after a good night of sleep. We shiver when we're cold and luxuriate in the warmth of a blanket. We feel friendship and caring and compassion and concern, and we use our intelligence to temper our lesser emotions. We enjoy gazing at the stars, listening to music, reading books, playing games, and embracing those we love. We should always strive to know more, but we should never lose sight of the gift of our very humanness.

References and Further Reading

1. Barrow, John. *The Constants of Nature: The Numbers That Encode the Deepest Secrets of the Universe*. New York: Vintage, 2009.

2. Boström, Nick. "Are You Living in a Computer Simulation?" *Philosophical Quarterly* 53, no. 211 (2003): 243–55. See also www.simulation-argument.com/simulation.html. Accessed July 11, 2017.

3. Dennett, Daniel. "Reflections on 'Free Will.'" Naturalism.org, 2014. naturalism.org/resources/book-reviews/reflections-on-free-will. Accessed July 11, 2017.

4. Deutsch, David. *The Fabric of Reality: The Science of Parallel Universes—and Its Implications*. New York: Penguin, 1997.

5. Graziano, Michael S. A. "Are We Really Conscious?" *New York Times Sunday Review*, October 10, 2014. See also www.nytimes.com/2014/10/12/opinion/sunday/are-we-really-conscious.html?_r=0. Accessed July 11, 2017.

6. Harris, Sam. *Free Will*. Washington, DC: Free Press, 2012.

Index